Bruce D Hopkins

HIGH RESOLUTION NMR IN THE SOLID STATE

High Resolution NMR in the Solid State

Fundamentals of CP/MAS

E. O. STEJSKAL
Professor of Chemistry
North Carolina State University
Raleigh

J. D. MEMORY
Vice President for Research
University of North Carolina
General Administration
        and
Professor of Physics
North Carolina State University
Raleigh

New York   Oxford
OXFORD UNIVERSITY PRESS
1994

Oxford University Press

Oxford   New York   Toronto
Delhi   Bombay   Calcutta   Madras   Karachi
Kuala Lumpur   Singapore   Hong Kong   Tokyo
Nairobi   Dar es Salaam   Cape Town
Melbourne   Auckland   Madrid

and associated companies in
Berlin   Ibadan

Library of Congress Cataloging-in-Publication Data
Stejskal, E. O., 1932–
High resolution NMR in the solid state: fundamentals of CP/MAS /
E.O. Stejskal and J.D. Memory.
p.   cm.   Includes bibliographical references and index.
ISBN 0-19-507380-0
1. Solids—Spectra.   2. Nuclear magnetic resonance spectroscopy.
3. Solid state chemistry.   I. Memory, Jasper D.   II. Title.
QC176.8.06S74   1995   543′.0887—dc20   94-6547

9 8 7 6 5 4 3 2 1

Printed in the United States of America
on acid-free paper

For
Anita Lasher Stejskal and Carolyn Hofler Memory

Preface

The aim of this book is to provide the NMR spectroscopist with a complete and detailed understanding of the cross-polarization (CP) magic-angle-spinning (MAS) experiment as it is used to obtain rare-spin, high-resolution NMR spectra. Most likely he or she will be primarily interested in ^{13}C spectra; but other spin-1/2 nuclei: ^{29}Si, ^{31}P, ^{15}N, etc. will behave in much the same way. NMR interactions in the solid state vary more widely than do the same interactions in solution. As a result, a detailed understanding of the CP/MAS experiment is necessary to establish the proper experimental conditions for the collection of a CP/MAS spectrum. Subsequently, the interpretation of spectra and relaxation measurements also requires an understanding of the interplay of several fundamental physical interactions.

The level of understanding we seek is one that will enable the spectroscopist to deal with unusual or difficult systems. If necessary, this will include the ability to design new pulse sequences to get at difficult spectral features or to focus on specific molecular motions or interactions. To do this the spectroscopist will have to develop a feel for the way interacting spin systems behave in the solid state.

There are two basically different ways to consider the interactions of spin systems. The first, and most popular among writers of high-level monographs, is the quantum mechanically rigorous approach. This approach is most useful for people who see things algebraically. The second way uses classical (or pseudo-classical) vector models to visualize how the spins behave, and only introduces quantum mechanical phenomena when necessary. In this approach the problem may be broken into separate parts, some of which behave classically and some of which do not. At the end, they are grafted together. The second way, although less rigorous, offers a way to visualize what is happening. It allows the spectroscopist to concentrate on the unique parts of his or her problem and to blend in later the parts that behave conventionally. It is our belief that many NMR innovators begin with their own version of this bastardized theory to assemble their ideas and then only make them rigorous after they know that the experiment will work. Furthermore, a truly rigorous approach is often computationally impossible. A certain qualitative visualization is necessary to develop approximations that will enable a rigorous treatment to be brought to a usable conclusion.

The first part of the book takes a historical point of view, since we believe that retracing the development of a subject is a good way to learn fundamentals. This is in keeping with the notion that the subject was originally built of individual, simple concepts and then made complete and rigorous later.

This book began as a set of lectures for physical chemistry and chemical physics graduate students specializing in CP/MAS NMR who wanted to understand the underlying principles of the experiments they were using in the laboratory, and is aimed at that kind of audience. Parts of the book will serve as reference to specific facts, such as how to set up a CP/MAS experiment for a particular class of systems, or how to use a Smith chart. Other parts are solely pedagogical in nature: once understood, never reread. Little time is spent on mathematical rigor, and appendices are provided to fill holes in student backgrounds.

Although some consideration is given to quadrupolar nuclei, particularly spin-1 nuclei, as simple examples of non-classical spin behavior, and also in the way that quadrupolar nuclei affect CP/MAS spectra, only passing attention will be paid to high-resolution experiments directly involving quadrupolar nuclei. Similarly, although I-I decoupling will be considered in detail in relation to rare-spin experiments, abundant-spin observed experiments such as CRAMPS will only be included for the sake of completeness.

E. O. S. would like to acknowledge countless hours spent discussing NMR phenomena with Jacob Schaefer in an attempt to develop a visualization of what is happening in the solid state of closely coupled spin systems. Many of the qualitative pictures in this book were the result of those discussions. He would also like to acknowledge many other discussions with other members of the New Analytical Techniques group of Monsanto Company, Corporate Research Department, especially Bob McKay, who showed him how to appreciate the Smith chart, and Tom Dixon, for his unusual insight into many NMR concepts. Both authors would also like to thank those students who listened and criticized as this particular approach was being developed, especially Jason Burgess, Shanmin Zhang, and Daniel Shin. The criticisms and suggestions of Z Z. Hugus, R. E. Fornes, and C. G. Moreland, as this material was first being organized, were also much appreciated. It was Ray Fornes who first suggested that we write this book. We are especially grateful to Tom Farrar, who shared his and John Harriman's excellent introductions to density matrix theory. This has affected materially how we have treated this subject. We are deeply grateful to Mrs. Shirley Maddry for her work in the preparation of the manuscript.

Contents

HIGH RESOLUTION NMR IN THE SOLID STATE

I

Introduction to NMR Principles

In this chapter, we will discuss the general principles of nuclear magnetic resonance (NMR), beginning with fundamental definitions and an introduction to the basic experiment, then writing down the Bloch Equations to describe the macroscopic nuclear magnetization and solving them for both continuous wave and pulse conditions. Then, after a brief review of relevant parts of quantum mechanics, including an introduction to density matrix theory, we will use the Solomon Equations to describe the effects of internuclear interactions, and then describe several useful pulse sequences.

I.A. FUNDAMENTAL DEFINITIONS

In this section, we will present the general physical principles of NMR at a basic level, first using a simple quantum mechanical approach, then from a classical perspective. Appendix A1 gives a review of mathematical principles that may be of some use in following this and subsequent chapters.

I.A.1. The Vector Model

We begin by considering a classical magnetic moment μ (the nuclear magnetic moment, in our case) in a magnetic field $\mathbf{B}_0$. This magnetic moment μ is related to angular momentum $\mathbf{L}$ by

$$\mu = \gamma \mathbf{L}, \tag{I.A.1}$$

which defines the magnetogyric ratio γ. It is only one step from Newton's second law to the equation

$$\frac{d\mathbf{L}}{dt} = \mathbf{N}, \tag{I.A.2}$$

which states that the time rate of change of angular momentum, $\mathbf{L} = \mathbf{r} \times m\mathbf{v}$, equals the torque, $\mathbf{N} = \mathbf{r} \times \mathbf{F}$. Using the fact that the torque $\mathbf{N}$ exerted on a magnetic moment μ in a magnetic field $\mathbf{B}_0$ is

$$\mathbf{N} = \mu \times \mathbf{B}_0, \tag{I.A.3}$$

and using Eqs. I.A.1–I.A.3, we obtain

$$\frac{d\mathbf{L}}{dt} = \boldsymbol{\mu} \times \mathbf{B}_0 = \gamma \mathbf{L} \times \mathbf{B}_0 = -\gamma \mathbf{B}_0 \times \mathbf{L}. \tag{I.A.4}$$

$\mathbf{B}_0$ is normally measured in Tesla (T), though one often sees $10^{-4}\,\mathrm{T} = 1$ gauss.

Comparison of Eqs. I.A.4 and A1.56 (see Appendix A1) leads us to conclude that a magnetic moment precesses about the direction of the magnetic field with an angular frequency

$$\omega_0 = \gamma \mathbf{B}_0 = 2\pi \mathbf{f}_0. \tag{I.A.5}$$

$\mathbf{f}_0$ is called the *Larmor precession frequency*, and is parallel to $\mathbf{B}_0$, but will be in the opposite direction if γ is positive. For instance, for protons (^{1}H) in a field of 1 Tesla (T), the resonant frequency is 42.57 MHz, or 2.675×10^8 radians/sec. It is common to drop the negative sign, and we will do so later when it no longer serves a purpose.

Though this discussion applies to a single magnetic dipole (spin), it is easily generalized to the macroscopic magnetization, which is defined as the total magnetic moment per unit volume:

$$\mathbf{M} = \sum_i \boldsymbol{\mu}_i, \tag{I.A.6}$$

where the sum includes all magnetic moments in a unit volume. Each step in the derivation of Eq. I.A.4 generalizes, so

$$\frac{d\mathbf{M}}{dt} = \mathbf{M} \times \gamma \mathbf{B}_0. \tag{I.A.7}$$

See Fig. I.A.1; $\mathbf{M}_\parallel$ and $\mathbf{M}_\perp$ are the parallel and perpendicular components of $\mathbf{M}$, as used later in Section I.B.2.

I.A.2. Simple Quantum Mechanical Description

We will write the intrinsic angular momentum of a nucleus as

$$\mathbf{L} = \hbar \mathbf{I}, \tag{I.A.8}$$

where $\hbar = h/2\pi$, Planck's constant divided by 2π, and where $\mathbf{I}$ is the spin angular momentum of the nucleus.

Quantum mechanics tells us that the magnitude of $\mathbf{I}$ is

$$|\mathbf{I}| = (\mathbf{I} \cdot \mathbf{I})^{1/2} = [I(I+1)]^{1/2} \neq I, \tag{I.A.9}$$

where the allowed values of I (the nuclear spin, a constant for a particular nucleus) are zero, a positive integer, or a postive half-integer: ^{12}C has spin 0, ^{13}C and ^{1}H have spin $\frac{1}{2}$, ^{14}N has spin 1, and so forth.

Associated with spin is a parallel magnetic moment $\boldsymbol{\mu}$, which is related to γ

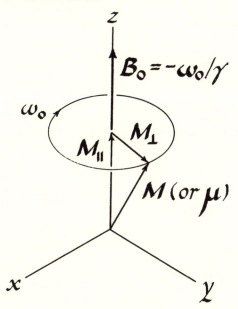

Figure I.A.1 Larmor precession.

and **I** through Eqs. I.A.1 and I.A.8:

$$\boldsymbol{\mu} = \gamma \hbar \mathbf{I}. \tag{I.A.10}$$

In a magnetic field $\mathbf{B}_0$ along the z-axis, μ has a potential energy given by

$$V = -\boldsymbol{\mu} \cdot \mathbf{B}_0 = -\hbar\gamma \mathbf{I} \cdot \mathbf{B}_0 = -\hbar\gamma I_z B_0 = -\hbar\gamma m B_0, \tag{I.A.11}$$

where the values allowed by quantum mechanics are

$$I_z = m, \, m = I, I - 1, \ldots, -1 \quad (2I + 1 \text{ values}). \tag{I.A.12}$$

For a nucleus of spin $\frac{1}{2}$, we have

$$I = \tfrac{1}{2}, \, m = \pm\tfrac{1}{2}, \, V_\pm = \mp\hbar\gamma B_0/2 \tag{I.A.13}$$

(see Fig. I.A.2).

The difference in energy between the two states of the nucleus (spin "up" and spin "down") is

$$\Delta V = V_- - V_+ = \hbar\gamma B_0 = hf_0 \text{ (or } h\nu_0\text{);} \tag{I.A.14}$$

$$V \uparrow \quad \begin{array}{ll} (m = -\tfrac{1}{2}) \,\rule[0.5ex]{4em}{0.4pt}\, & V_- = (\tfrac{1}{2})\hbar\gamma B_0 \\ 0 & \\ (m = +\tfrac{1}{2}) \,\rule[0.5ex]{4em}{0.4pt}\, & V_+ = -(\tfrac{1}{2})\hbar\gamma B_0 \end{array}$$

Figure I.A.2 Nuclear energy levels ($I = 1/2$).

f_0 (or v_0) is the frequency of radiation required to cause transitions between the two levels; such a transition is called *nuclear magnetic resonance* (NMR). One often uses the angular frequency

$$\omega_0 = 2\pi f_0 = \gamma B_0. \tag{I.A.15}$$

Quantum theory states that electromagnetic radiation satisfying Eq. I.A.15 will stimulate transitions *both* ways: from "down" to "up" and *vice versa*. There is usually a net absorption, because typically there are more nuclei in the lower level than the upper level.

I.A.3. The Boltzmann Distribution and Curie Susceptibility

For a macroscopic sample, the ratio of the populations of the two levels at an absolute temperature T is governed by the Boltzmann factor:

$$N(-\tfrac{1}{2})/N(+\tfrac{1}{2}) = N_-/N_+$$

$$= \exp\left\{-\frac{\tfrac{1}{2}\hbar\omega_0}{kT}\right\}\Big/\exp\left\{-\frac{-\tfrac{1}{2}\hbar\omega_0}{kT}\right\} = \exp\left\{-\frac{\hbar\omega_0}{kT}\right\} = e^{-\Delta V/kT}. \tag{I.A.16}$$

At normal temperatures $\hbar\omega_0 \ll kT$, so the exponential can be expanded and only two terms kept:

$$N_-/N_+ \approx 1 - \frac{\hbar\omega_0}{kT}. \tag{I.A.17}$$

It is simple to derive from this the Curie law for paramagnetic susceptibility of the nuclear spins. The susceptibility χ_0 is defined as the constant of proportionality between the external magnetic field B_0 and the induced magnetization M:

$$M = \chi_0 B_0. \tag{I.A.18}$$

But the induced magnetization is just the fractional excess of nuclei in the spin "up" state, times the magnetic moment $\hbar\gamma I_z$, times the total number of nuclei per unit volume n:

$$M = n\hbar\gamma I_z\left(\frac{N_+ - N_-}{N_+ + N_-}\right) = n\hbar\gamma(\tfrac{1}{2})\left[\frac{1 + \hbar\gamma B_0/2kT - 1 + \hbar\gamma B_0/2kT}{1 + \hbar\gamma B_0/2kT + 1 - \hbar\gamma B_0/2kT}\right]$$

$$= \tfrac{1}{4}n\hbar^2\gamma^2 B_0/kT, \tag{I.A.19}$$

so that

$$\chi_0 = n\hbar^2\gamma^2/4kT. \tag{I.A.20}$$

This inverse dependence of the susceptibility on the absolute temperature is the Curie law of paramagnetism.

I.A.4. The Spin Temperature

Even though the Boltzmann relation (Eq. I.A.16) applies strictly only for a system in thermal equilibrium, it is sometimes appropriate to define a "spin temperature" T_s in terms of the instantaneous populations of the spin states:

$$N_-/N_+ = e^{-\Delta V/kT_s}. \tag{I.A.21}$$

At the beginning of an experiment, radiation whose frequency satisfies Eq. I.A.15 will result in a net absorption, since there are more spins in the lower energy state to go up to than there are in the higher states to go down. After a number of transitions have taken place, N_- will have increased and N_+ will have decreased so the ratio in Eq. I.A.16 will increase, approaching unity. This corresponds, through Eq. I.A.21, to an *increase* in spin temperature T_s, a "warming up" of the spin system. If the populations were to become equal, further irradiation will produce no *net* absorption, and the system will be said to be "saturated". In this brief discussion, we have omitted other processes that will work to reestablish the thermal equilibrium populations. We will return to this matter, nuclear magnetic relaxation, in Section I.B.1.

I.A.5. Dipolar Broadening, Chemical Shielding, and Spin–Spin Coupling

The nuclear magnetic moments of nuclei such as ^{13}C and ^1H, which make possible the observation of NMR, have an additional effect: they produce small, localized magnetic fields, themselves, at other nuclear sites. Since the resonance frequency depends on the local field at the site of the resonating nucleus, and since the local magnetic field will vary from point to point throughout the sample due to different orientations of neighboring nuclei, one gets a broadened NMR line for solid samples. This broadening, which arises due to the interactions of the nuclear dipoles, is called *dipole–dipole* broadening. The local field typically varies over a range of several mT. How this broadening can be reduced experimentally will be a major topic in later chapters.

With liquid samples, rapid and random rotation of molecules in the sample causes this broadening to disappear, owing to the averaging of the dipole–dipole interaction, resulting in narrow lines. Under such high resolution conditions, it is possible to observe hyperfine structure in NMR spectra: *chemical shielding* and *spin–spin coupling*.

The early NMR physicists viewed nuclear spins suspended or in a vacuum, and hence used H instead of B. In the presence of matter, chemical shielding, the "physicist's disappointment" (and chemist's delight), requires the use of B. To complicate matters further, in the old gaussian unit system, $H = B$ in a vacuum, although they are measured in gauss and oersted respectively, units which we found in many books and articles. One Tesla = 10^4 oersted.

The presence of electrons in the sample containing the nucleus we have been discussing makes itself felt in chemical shielding. Putting the sample in the strong, constant external field $\mathbf{B}_0$ induces electronic currents that produce, in turn, an induced magnetic field $\mathbf{B}'$ that modifies the local magnetic field $\mathbf{B}_{loc}$ at the site of the nucleus in question. $\mathbf{B}'$ will be proportional to $\mathbf{B}_0$, and generally diamagnetic; one defines the chemical shielding parameter σ by the equation

$$B_{loc} = B_0(1 - \sigma). \tag{I.A.22}$$

The absorption frequency of the nucleus, then, is

$$f = \gamma B_{loc}/2\pi = (\gamma/2\pi)B_0(1 - \sigma) = f_0(1 - \sigma), \tag{I.A.23}$$

where we have used Eqs. I.A.15 and I.A.22.

Nuclei in different chemical environments will experience different local fields, and hence will resonate at different frequencies. One defines a specific reference frequency, typically that of tetramethylsilane (TMS):

$$f_{ref} = f_0(1 - \sigma_{ref}), \tag{I.A.24}$$

so that we can define a chemical shift δ by

$$\delta = 10^6 \frac{f - f_{ref}}{f_0}, \tag{I.A.25}$$

which is expressed in parts per million (ppm).

A more precise definition which does not depend on the spectrometer frequency is $10^6(f - f_{ref}/f_{ref}) = \delta$; however, the difference is extremely small, and Eq. I.A.25 is commonly used.

Note that, from Eqs. I.A.22–I.A.25, δ is *independent* of B_0. We also note here that in plotting an NMR spectrum, it is traditional to have δ increasing to the left.

In a typical sample, there will be nuclei in more than one chemical environment, hence more than one σ, hence more than one NMR frequency. Consider two environments labeled A and X characterized by ω_A and ω_X:

$$\left. \begin{array}{l} \omega_A = \gamma_A B_0(1 - \sigma_A) \\ \omega_X = \gamma_X B_0(1 - \sigma_X). \end{array} \right\} \tag{I.A.26}$$

The energy of interaction with an external field is

$$V(m_A m_A) = -\hbar\omega_A m_A - \hbar\omega_X m_X, \tag{I.A.27}$$

where possible values are

$$m_A, m_X = \pm\tfrac{1}{2}, \tag{I.A.28}$$

(see Fig. I.A.3).

Quantum mechanics prescribes that allowed transitions correspond to $\Delta m = \pm 1$:

$$\left. \begin{array}{l} X: \Delta m_X = \pm 1, \\ \Delta m_A = 0; \\ A: \Delta m_A = \pm 1, \\ \Delta m_X = 0. \end{array} \right\} \tag{I.A.29}$$

So that there are two frequencies in the NMR spectrum, ω_A and ω_X.

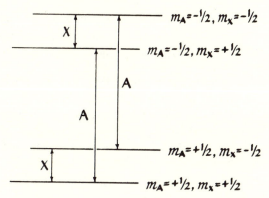

Figure I.A.3 Energy level diagram for an A–X system.

There may also be an interaction between the nuclei themselves arising from an indirect coupling through the electron spins (the Fermi contact hyperfine interaction) called the spin–spin interaction, or the J interaction. This modifies the energy thus:

$$V(m_A, m_X) = -\hbar\omega_A m_A - \hbar\omega_X m_X + hJ_{AX}m_A m_X, \qquad \text{(I.A.30)}$$

which causes a further splitting in the energy level system, since the last term is positive or negative depending on the values of m_A and m_X. Figures I.A.4 and I.A.5 show the new energy level diagram and the resulting NMR spectrum. Reference to Eqs. I.A.26 and I.A.30 reminds us that the ωs are B_0 dependent, but the J splittings are not. J is always expressed in Hertz, hence the h in Eq. I.A.30.

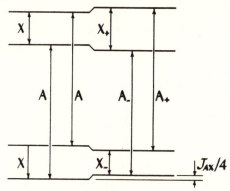

Figure I.A.4 Energy level diagram for an A–X system, add J_{AX}.

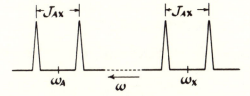

Figure I.A.5 A–X NMR spectrum.

In most systems, more than two spins will interact through spin–spin coupling, which gives rise to a more complicated spectrum. The experimeter can simplify a spectrum through *decoupling*, a procedure in which one resonance frequency, say ω_X, is strongly irradiated, which collapses the doublet in the ω_A resonance due to the J coupling between A and X. A rough way of explaining this is to say that under decoupling conditions, the X nucleus is making rapid transitions between spin up and spin down states, thus averaging to zero the effect of the J coupling on A. There are other interactions in the solid state that we will consider later, that are not usually seen in liquid state NMR, where molecules tumble freely.

I.A.6. RF Radiation and the Rotating Frame

We now turn to an alternative, classical approach to the NMR experiment. We have, from Eq. I.A.4, a magnetic moment μ precessing about the direction of an external magnetic field $\mathbf{B}_0$ along the z-axis with angular velocity ω. The NMR experiment includes a small, time-dependent, additional magnetic field $\mathbf{B}_1$, rotating in the x–y plane in the same sense and with the same angular velocity as μ. $\mathbf{B}_1$ is taken to be along the x'-axis in the rotation frame. In practice, one uses a linearly polarized oscillating field, but this can be resolved into two components rotating with opposite senses. Only that component rotating in the same sense as μ has any significant effect on it; the other can be disregarded. In the presence of both $\mathbf{B}_0$ and $\mathbf{B}_1$, μ obeys the equation

$$\frac{d\lambda}{dt} = \gamma\mu \times (\mathbf{B}_0 + \mathbf{B}_1). \tag{I.A.31}$$

We now transform to a coordinate system rotating about the z-axis with angular velocity ω in which $\mathbf{B}_1$ will stand still, using the methods described in Appendix A1:

$$\frac{d\mu}{dt} = \frac{d\mu'}{dt} + \omega \times \mu'. \tag{I.A.32}$$

Using Eq. I.A.31, we have

$$\gamma\mu' \times (\mathbf{B}_0 + \mathbf{B}_1) = \frac{d\mu'}{dt} + \omega \times \mu', \tag{I.A.33}$$

so that

$$\frac{d\mu'}{dt} = \omega'_0 \times \mu', \tag{I.A.34}$$

where

$$\omega'_0 = -(\omega + \gamma\mathbf{B}_0 + \gamma\mathbf{B}_1). \tag{I.A.35}$$

In the rotating frame (see Fig. I.A.6), then, the equation of motion of the dipole is

$$\frac{d\mu'}{dt} = \gamma\mu' \times \mathbf{B}_{\text{eff}}, \tag{I.A.36}$$

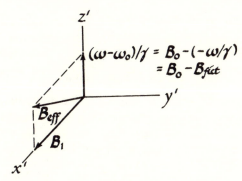

Figure I.A.6 Effective field in the rotating frame.

where we have written

$$\gamma \mathbf{B}_{\mathrm{eff}} = (\omega - \omega_0) + \gamma \mathbf{B}_1 = -\omega_{\mathrm{eff}}. \tag{I.A.37}$$

For $\omega = \omega_0$ and $B_1 = 0$, μ' will be stationary in the rotating frame. Many of our later calculations will be simplified by the use of the rotating frame.

I.B. THE MACROSCOPIC NUCLEAR MAGNETIZATION AND THE BLOCH EQUATIONS

In this section, we will introduce nuclear magnetic relaxation, and use the results of section I.A.6 to establish the Bloch Equations for the time-dependence of the macroscopic nuclear magnetization. We will then solve the equations for a variety of important special cases; for example, pulse experiments on the one hand and continuous wave experiments on the other.

I.B.1. The Relaxation Times T_1 and T_2

There are processes that lead to the restoration of the thermal equilibrium populations of the spin states in the case that those populations are disturbed. Such a case would occur after the absorption of radiation at the proper NMR frequency, for example. The processes involve the interaction of the spin system with the rest of the "lattice" or "bath" comprising the sample. This *nuclear magnetic relaxation* can be viewed as a reapproach of the spin temperature T_s to the temperature of the lattice T.

The *relaxation time* is a measure of the speed with which T_s approaches T. Under many circumstances, there are two different relaxation times: T_1, the *longitudinal* or *spin-lattice* relaxation time, which measures the rate at which the component of the magnetization parallel to the external magnetic field $\mathbf{B}_0$ approaches its thermal equilibrium value, which involves processes in which there is an exchange of energy between the spin system and the lattice; and T_2, the *transverse*, or *spin–spin* relaxation time, which measures the rate at which the magnetization in the plane normal to $\mathbf{B}_0$ approaches its return to equilibrium. Since processes leading to this do not require a change in the energy of the spin system, it is clear that T_1 and T_2 are not necessarily equal.

I.B.2. The Bloch Equations and the Rotating Frame

With these definitions, we are prepared to write down the Bloch Equation:

$$\frac{d\mathbf{M}}{dt} = \gamma \mathbf{M} \times \mathbf{B} - \frac{\mathbf{M}_\perp}{T_2} + \frac{(\mathbf{M}_0 - \mathbf{M}_\parallel)}{T_1}. \tag{I.B.1}$$

(Since this is a *vector* equation, it is sometimes written in component form, in which we refer to the Bloch Equation*s*).

The first term would describe the precession of $\mathbf{M}$ about $\mathbf{B}$, but at equilibrium, $\mathbf{M}$ is parallel to $\mathbf{B}$ and this term drops out.

We will give a brief justification for this term by term. In equilibrium, (that is, when $d\mathbf{M}/dt = 0$),

$$\mathbf{M}_\parallel^{eq} = \mathbf{M}_0 \tag{I.B.2}$$

and

$$\mathbf{M}_\perp^{eq} = 0; \tag{I.B.3}$$

that is, taking $\mathbf{B}$ along the z-axis, the z component of $\mathbf{M}$ has its equilibrium value $\mathbf{M}_0$, and the component at right angles to the z-axis is zero. If $\mathbf{M}_\perp \neq 0$, $\mathbf{M}_\perp$ will decrease at a rate inversely proportional to T_2; if $\mathbf{M}_\parallel \neq \mathbf{M}_0$, $\mathbf{M}_\parallel$ will change at a rate inversely proportional to T_1.

We will be solving the Bloch Equation for a variety of experimental situations. First, however, let us consider how it looks in the rotating frame (see Appendix A1 and Section I.A.6). We have

$$\frac{d\mathbf{M}}{dt} = \frac{d\mathbf{M}'}{dt} + \omega \times \mathbf{M}. \tag{I.B.4}$$

Using Eqs. I.B.1 and I.B.4, we obtain

$$\frac{d\mathbf{M}'}{dt} = \mathbf{M} \times (\omega - \omega_0 + \gamma\mathbf{B}_1) - \frac{\mathbf{M}_\perp}{T_2} + \frac{(\mathbf{M}_0 - \mathbf{M}_\parallel)}{T_1}. \tag{I.B.5}$$

I.B.3. The Slow Passage Continuous Wave Solution – Saturation

We now apply the Bloch Equation to the original, continuous wave NMR experiment. Instead of a pulse (which we will treat, presently) of radiation at the resonance frequency and at high B_1, one irradiates continuously at a *small B_1* and varies the angular frequency (or the value of B_0) slowly to obtain the spectrum. Bloch's equation in the rotating frame and in the steady state (that is, for $d\mathbf{M}'/dt = 0$) is

$$\frac{d\mathbf{M}'}{dt} = \mathbf{M} \times (\omega - \omega_0 + \gamma\mathbf{B}_1) - \frac{\mathbf{M}_\perp}{T_2} + \frac{(\mathbf{M}_0 - \mathbf{M}_\parallel)}{T_1} = 0. \tag{I.B.6}$$

We write

$$\mathbf{M} = \mathbf{M}_\parallel + \mathbf{M}_\perp = \mathbf{M}_\parallel + \mathbf{M}_u + \mathbf{M}_v, \qquad \text{(I.B.7)}$$

where we have along the z' axis:

$$\omega_0 = \gamma \mathbf{B}_0, \ \mathbf{M}_0, \ \omega, \ \mathbf{M}_\parallel, \ \mathbf{k}; \qquad \text{(I.B.8)}$$

along the x'-axis:

$$\mathbf{B}_1, \ \omega_1 = \gamma \mathbf{B}_1, \ \mathbf{M}_u, \ \mathbf{i}'; \qquad \text{(I.B.9)}$$

and along the y'-axis:

$$\mathbf{M}_v, \mathbf{j}'. \qquad \text{(I.B.10)}$$

Using the vector method outlined in Appendix A1 and straightforward algebra, we obtain

$$M_u = \frac{M_0 \gamma B_1 T_2^2 (\omega - \omega_0)}{1 + T_2^2 (\omega_0 - \omega)^2 + \gamma^2 B_1^2 T_1 T_2} \qquad \text{(I.B.11)}$$

$$M_v = \frac{M_0 \gamma B_1 T_2}{1 + T_2^2 (\omega_0 - \omega)^2 + \gamma^2 B_1^2 T_1 T_2} \qquad \text{(I.B.12)}$$

$$M_\parallel = \frac{M_0 [1 + T_2^2 (\omega_0 - \omega)^2]}{1 + T_2^2 (\omega_0 - \omega)^2 + \gamma^2 B_1^2 T_1 T_2} \qquad \text{(I.B.13)}$$

for steady state conditions. One detects experimentally M_u and/or M_v as a function of ω. Recall

$$V = -\mu \cdot \mathbf{B}_0, \qquad \frac{d\mathrm{V}}{dt} = -\frac{d\mu}{dt} \cdot \mathbf{B}_0, \qquad \text{(I.B.14)}$$

Since

$$\frac{d\mathbf{\mu}}{dt} = \gamma \mathbf{\mu} \times (\mathbf{B}_1 + \mathbf{B}_0), \qquad \text{(I.B.15)}$$

we have

$$\frac{dV}{dt} = -\gamma \mathbf{\mu} \times (\mathbf{B}_1 + \mathbf{B}_0) \cdot \mathbf{B}_0 = -\gamma \mathbf{\mu} \times \mathbf{B}_1 \cdot \mathbf{B}_0. \qquad \text{(I.B.16)}$$

Since

$$\mathbf{M} = \sum_i \mathbf{\mu}_i, \qquad \text{(I.B.17)}$$

we obtain, for the rate of energy absorption,

$$A = -\gamma \mathbf{M} \times \mathbf{B}_1 \cdot \mathbf{B}_0 = +\gamma M_v \mathbf{B}_1 \cdot \mathbf{B}_0, \qquad \text{(I.B.18)}$$

where we have used

$$\mathbf{j} \cdot (\mathbf{i} \times \mathbf{k}) = -1. \qquad \text{(I.B.19)}$$

Or, using Eq. I.B.19,

$$A = \chi_0 \frac{(\omega_0 B_1)^2 T_2}{1 + T_2^2 (\omega_0 - \omega)^2 + \gamma^2 B_1^2 T_1 T_2}, \qquad \text{(I.B.20)}$$

where χ_0, the static susceptibility, is defined in Eq. I.A.18.

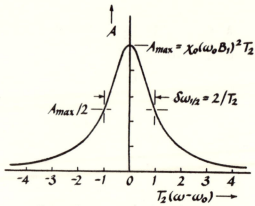

Figure I.B.1 Lorentzian line (absorption).

We define the saturation factor s to be the last term in the denominator:

$$s = \gamma^2 B_1^2 T_1 T_2. \qquad (I.B.21)$$

Saturation occurs when the populations of the spin states are the same, under which condition further net energy absorption stops. Note that the saturation factor s is proportional to B_1^2, and the relaxation times. This is plausible since a large B_1 means increased radiative transition probability, which leads to an equalization of spin populations, and long relaxation times mean slow action in restoring thermal equilibrium populations. If $s \ll 1$, Eq. I.B.20 becomes

$$A = \chi_0 \frac{(\omega_0 B_1) T_2}{1 + T_2^2 (\omega_0 + \omega)^2}, \qquad (I.B.22)$$

which is a *Lorentzian* line (see Fig. I.B.1).

The half-maximum points are found by

$$A_m/2 = A_m / [1 + T_2^2 (\omega_0 - \omega)^2]. \qquad (I.B.23)$$

This leads to $\pm(\omega_0 - \omega) = 1/T_2$. The spectrum width at half-maximum, then, is

$$\delta\omega_{1/2} = \frac{2}{T_2}. \qquad (I.B.24)$$

The other component, M_u, is detected as the *dispersion* (see Fig. I.B.2).

From Eq. I.B.13, we have

$$M_{\parallel} = \frac{\chi_0 B_0 [1 + T_2^2 (\omega_0 - \omega)^2]}{1 + T_2^2 (\omega_0 - \omega)^2 + \gamma^2 B_1^2 T_1 T_2}. \qquad (I.B.25)$$

When $\omega_0 - \omega = 0$ (on resonance), this becomes

$$M_{\parallel} = \frac{\chi_0 B_0}{1 + \gamma^2 B_1^2 T_1 T_2} = \frac{\chi_0 B_0}{1 + s}. \qquad (I.B.26)$$

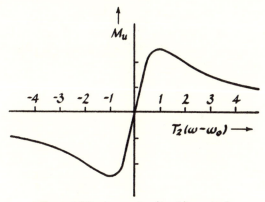

Figure I.B.2   Lorentzian line (dispersion).

Also, Eq. I.B.20 becomes

$$A = \chi_0 \frac{\omega_0^2 B_1^2 T_2}{1 + T_2^2(\omega_0 - \omega)^2 + s}. \tag{I.B.27}$$

At resonance,

$$A = A_{max} = \chi_0 \frac{\omega_0^2 B_1^2 T_2}{1 + s}, \tag{I.B.28}$$

so we can write

$$A = A_{max} \left\{ \frac{1 + s}{1 + s + T_2^2(\omega_0 - \omega)^2} \right\}. \tag{I.B.29}$$

Now we can ask again what the width of the absorption curve is at the half-maximum:

$$\omega_0 - \omega = \pm \sqrt{1 + s}/T_2. \tag{I.B.30}$$

So now we have

$$\delta\omega_{1/2} = \frac{2\sqrt{1 + s}}{T_2}. \tag{I.B.31}$$

This increase in linewidth (compare with Eq. I.B.24) is called *saturation broadening*.

I.B.4. Pulsed NMR

Now consider the effect of a pulse of magnitude B_1 at frequency ω_0 (so that B_1 is stationary in the rotating frame), but short enough with respect to T_1 and T_2 so that the relaxation terms can be neglected. The magnitude of B_1 should be large relative to that used in the continuous wave experiment. Then

$$\frac{d\mathbf{M}'}{dt} = \mathbf{M} \times (\gamma \mathbf{B}_{eff}), \tag{I.B.32}$$

where we have defined an effective magnetic field $\mathbf{B}_{eff}$ in Eq. I.A.30.

According to Eq. I.B.32, $\mathbf{M}$ precesses in the rotating frame about $\mathbf{B}_{\text{eff}}$ (see Fig. I.A.6). At resonance, that is if $\omega = \omega_0$, $\mathbf{M}$ will precess with unchanged length in the z–y' plane, about $\mathbf{B}_1$, which is stationary along the x'-axis in the rotating frame. The angular velocity of precession, from Eq. I.A.30, is

$$\omega_1 = \gamma B_1. \tag{I.B.33}$$

If the duration of the pulse is t_p, then the magnetization will have precessed by an angle

$$\theta = \omega_1 t_p = \gamma B_1 t_p. \tag{I.B.34}$$

Two special types of pulses are the $\pi/2$ pulse and the π pulse, for which θ is, respectively, $\pi/2$ and π. Use of Eq. I.B.34 tells us that for protons, given a value of 1 mT for B_1, we have $t_{\pi/2} = 6$ µs, for example.

What happens after a $\pi/2$ pulse? Returning to Eq. I.B.5. we have the first term vanishing, so

$$\frac{d\mathbf{M'}}{dt} = -\frac{\mathbf{M}_\perp}{T_2} = \frac{(\mathbf{M}_0 - \mathbf{M}_\parallel)}{T_1}. \tag{I.B.35}$$

Note that

$$M_z = M'_z = |M_\parallel| \tag{I.B.36}$$

and

$$\mathbf{M}_\perp = M'_x \mathbf{i'} + M'_y \mathbf{j'}. \tag{I.B.37}$$

Writing down the y'-component of Eq. I.B.35 gives

$$\frac{dM'_y}{dt} = -\frac{M'_y}{T_2}, \tag{I.B.38}$$

which is solved, as can be shown by substitution, by

$$M'_y = M_0 \, e^{-t/T_2}. \tag{I.B.39}$$

The z-component is

$$\frac{dM'_z}{dt} = \frac{M_0 - M_z}{T_1}, \tag{I.B.40}$$

solved by

$$M'_z = M_z = M_0(1 - e^{-t/T_1}). \tag{I.B.41}$$

We will now make use of the similarity between the mathematics of vectors in the x–y plane and the mathematics of complex numbers (see Appendix A1) in solving the Bloch Equation. Consider the case in which a $\pi/2$ pulse has just rotated the magnetization to the x-axis so that $\mathbf{M} = M_0 \mathbf{i'}$. From Eq. I.B.5, we have

$$\frac{d\mathbf{M}_\perp}{dt} = \gamma \mathbf{M}_\perp \times \mathbf{B}_0 - \frac{\mathbf{M}_\perp}{T_2} \tag{I.B.42}$$

and

$$\frac{d\mathbf{M}_\parallel}{dt} = \frac{(\mathbf{M}_0 - \mathbf{M}_\parallel)}{T_2} \tag{I.B.43}$$

Eq. I.B.42 leads to

$$\frac{dM_u \mathbf{i}}{dt} + \frac{dM_v \mathbf{j}}{dt} = \gamma(N_v \mathbf{i} - M_u \mathbf{j})B_0 - (M_u \mathbf{i} + M_v \mathbf{j})/T_2, \qquad \text{(I.B.44)}$$

where we have used Eq. I.B.7.

Now use complex variables for $\mathbf{M}_\perp$:

$$m = M_u + iM_v. \qquad \text{(I.B.45)}$$

Eq. I.B.51 becomes

$$dm/dt = \gamma H_0(-im) - m/T_2,$$
$$= -i\omega_0 m - m/T_2, \qquad \text{(I.B.46)}$$

since

$$im = iM_u - M_v. \qquad \text{(I.B.47)}$$

We can show by substitution that a solution to Eq. I.B.46 is

$$m = m_0 \exp\{-[i\omega_0 + (1/T_2)]t\}. \qquad \text{(I.B.48)}$$

If we write this as

$$m = m_0 \, e^{-t/T_2} \, e^{-i\omega_0 t}, \qquad \text{(I.B.49)}$$

we see that we have both *precession* and *decay*.

Now suppose B_0 varies over the sample, so that there is a part of the sample with

$$\omega = \omega_0 + \delta\omega. \qquad \text{(I.B.50)}$$

For this part, Eq. I.B.46 becomes

$$dm/dt = -i(\omega_0 + \delta\omega)m - m/T_2. \qquad \text{(I.B.51)}$$

We will solve this by letting m_0 vary with time:

$$dm/dt = (dm_0/dt) \exp\{-[i\omega_0 + (1/T_2)]t\}$$
$$+ m_0\{-[i\omega_0 + (1/T_2)]\} \exp\{-[i\omega_0 + (1/T_2)]t\}$$
$$= -[i(\omega_0 + \delta\omega) + (1/T_2)]m_0 \exp\{-[i\omega_0 + (1/T_2)]t\}, \qquad \text{(I.B.52)}$$

which simplifies to:

$$dm_0/dt = -i\,\delta\omega\,m_0. \qquad \text{(I.B.53)}$$

We can show by substitution that this is solved by

$$m_0 = A \exp\{-i\,\delta\omega\,t\} = A\,e^{-i\delta\omega t}, \qquad \text{(I.B.54)}$$

where

$$m_0 = A \quad \text{at } t = 0, \qquad \text{(I.B.55)}$$

so that

$$m = A \exp\{-i\delta\omega t\} \exp\{-[i\omega_0 + (1/T_2)]t\}. \qquad \text{(I.B.56)}$$

In general,

$$m_0 = A\,e^{-i(\delta\omega t + \varphi)}. \qquad \text{(I.B.57)}$$

A single pulse usually leaves all parts of the sample with the same φ, so, without loss of generality, we let it be 0.

We now integrate over the entire sample distribution, obtaining

$$M_0(t) = \int G(\omega - \omega_0) \, e^{-i(\omega - \omega_0)t} \, d\omega, \tag{I.B.58}$$

where $G(\omega - \omega_0)$ is a distribution function for the precession frequencies of nuclei over the macroscopic sample, which satisfies the normalization condition

$$\int G(\omega - \omega_0) \, d\omega = 1. \tag{I.B.59}$$

If we measure ω relative to ω_0, we can rewrite this as

$$M_0(t) = \int G(\omega) \, e^{-i\omega t} \, d\omega. \tag{I.B.60}$$

This looks like a continuous, complex Fourier transform (FT). In the most basic pulsed NMR experiment, one detect's the magnetization's free precession and decay in the x–y plane following a $\pi/2$ pulse (also note Eq. I.B.49 and subsequent comment). The observed signal is called a free induction decay (FID), in which the nuclear induction signal arising from the free precession following a $\pi/2$ pulse decays owing to the spread in precession frequency. An inverse FT can turn M_0 back into $G(\omega)$, the NMR spectrum.

The actual "signal", of course, will have the factor $\exp[-t/T_2]$, leading to a decay, sometimes called the Bloch decay, and sometimes called the free-induction decay (FID).

I.B.5. Some Special Pulse Sequences

Fig. I.B.3 shows a variation of the famous Hahn *spin echo* experiment: a $\pi/2$ pulse produces the usual FID, followed by a π pulse, which refocuses the spins which have spread out due to slightly different precession frequencies, with the result that there is another nuclear induction signal – an echo. The mathematics is as follows: start with $\pi/2$ pulse along the $-i$ axis at time $t = 0$. After a time t, we have,

$$m_0 = A \, e^{-i\delta\omega t}. \tag{I.B.61}$$

At $t = \tau$ apply a π pulse along the positive real axis. Following that, we have

$$m_0 = A \, e^{+i\delta\omega\tau} \, e^{-i\delta\omega(t-\tau)}$$

$$= A \, e^{-i[\delta\omega(t-2\tau)]}, \tag{I.B.62}$$

so that at $t = 2\tau$, $m_0 = A$, and we are right back where we started. If we sum over the entire sample (see M_0, above) we get a spin echo. The height of the echo will be somewhat less than the maximum value of the FID. This is due to spin–spin relaxation, in contrast to the dephasing that occurs in the original FID, which is typically due to magnetic field inhomogeneity across the sample. One can measure T_2 by repeating the spin echo experiment a number of times, in which the timing of the π pulse is varied (see Fig. I.B.3).

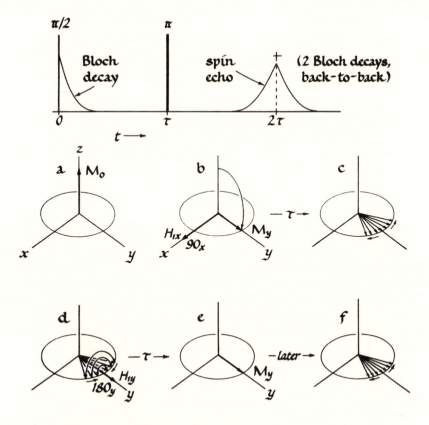

Hahn Spin-Echo Sequence

Figure I.B.3 Hahn spin-echo sequence.

An alternative method removes the necessity of multiple experiments and is also better in case of rapid molecular diffusion (see Section I.B.7) in the sample: one uses a *set* of π pulses, repeatedly refocusing the spins. Observations of the declining maxima of the successive pulses leads to a determination of T_2. A further refinement involves shifting the phase of the π pulses by 90°, which reduces the effect of pulse imperfections (see Chapter V).

Spin echo techniques can be helpful in measuring T_1. In the normal inversion recovery experiment, a π pulse inverts the nuclear magnetization. This will decay along the z-axis with the time constant T_1 (see Fig. I.B.4). Some time τ after the original π pulse, one flips the magnetization into the x–y plane and observes the extent of recovery to produce an echo, the amplitude of which will mirror the z-magnetization at the time τ. The experiment is repeated varying τ, to determine the decay back to the thermal equilibrium value of M_z, and hence measure T_1. In a very inhomogeneous field where it may be difficult to observe an FID following the $\pi/2$ pulse because of a short T_2^*, a spin echo may be created by following the $\pi/2$ pulse with another π pulse.

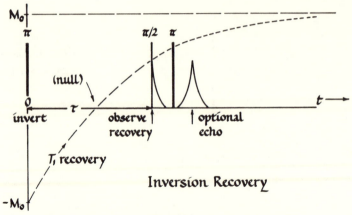

Figure I.B.4 T_1 measurement by inversion recovery.

Another frequently useful NMR pulse procedure is *spin-locking* (to be discussed in more detail in Section II.D). One uses a $\pi/2$ pulse to rotate the nuclear magnetization into the x–y plane. Assume that the pulse was along the x-axis so that the magnetization now lies along the y-axis in the rotating frame. One then changes the phase of the oscillating magnetic field so that it now lies along the y-axis collinear with the spin magnetization. The oscillating magnetic field is changed to B_{1y} along the y-axis in the rotating frame, and is maintained at that value for the duration of the spin lock process. The first two steps are indicated diagrammatically as Fig. II.D.2. The magnetic field seen by the spin magnetization is illustrated in Fig. II.D.1. The magnetization along the y-axis and the locking field appear constant in the rotating frame. The magnetization sees an effective field given by

$$B_{eff} = [(B_0 - \omega/\gamma)^2 + B_{1y}^2]^{1/2}, \tag{I.B.63}$$

where

$$\omega_0 = \gamma B_0. \tag{I.B.64}$$

In the rotating frame, the magnetization will precess about B_{eff} with an angular frequency given by

$$\omega_1 = \gamma B_{eff}. \tag{I.B.65}$$

For

$$\omega = \omega_0, \tag{I.B.66}$$

$$\omega_1 = \gamma B_{1y}. \tag{I.B.67}$$

It is assumed that the oscillating magnetic field B_{1y} is much less than the strong constant external magnetic field B_0, but much larger than the local field due to neighboring nuclei.

The spin lock procedure is central to the cross polarization process described in Chapter II.D.

We know that the sequence of pulses $\pi/2 - \tau - \pi$ produces an echo at 2τ; how about the more general sequence $\pi/2 - \tau - \theta$ (a $\pi/2$ pulse followed by an evolution time τ followed by a pulse θ of general length). Consider magnetization precessing in the rotating complex plane with frequency $\delta\omega$ after a $\pi/2$ pulse at $t = 0$ (see Eq. I.B.61). At time $t = \tau$ apply a θ pulse along the positive real axis (we are only concerned about the behavior in the u–v plane).

Before the pulses, we have

$$m_0 = M_0\, e^{-i\delta\omega\tau}$$

$$= M_0\{\cos \delta\omega\,\tau - i\sin \delta\omega\,\tau\}; \tag{I.B.68}$$

after the pulse (instantaneously), this becomes

$$m_0 = M_0\{\cos \delta\omega\,\tau - i(\cos \theta)\sin \delta\omega\,\tau\}$$

$$= M_0\left\{\frac{e^{+i\delta\omega\tau} + e^{-i\delta\omega\tau}}{2} - i(\cos \theta)\frac{e^{+i\delta\omega\tau} - e^{-i\delta\omega\tau}}{2i}\right\}$$

$$= \frac{M_0}{2}(1 - \cos \theta)e^{+i\delta\omega\tau} + \frac{M_0}{2}(1 + \cos \theta)e^{-i\delta\omega\tau}. \tag{I.B.69}$$

(see Appendix A1).

Note that the first term has had its phase inverted, but the second term is unchanged. This is now the starting point of a new evolution beginning at $t = \tau$:

$$m_0 = \left\{\frac{M_0}{2}(1 - \cos \theta)e^{+i\delta\omega\tau} + \frac{M_0}{2}(1 + \cos \theta)e^{-i\delta\omega(t-\tau)}\right.$$

$$= \frac{M_0(1 - \cos \theta)}{2}e^{-i\delta\omega\tau(t-2\tau)} + \frac{M_0(1 + \cos \theta)}{2}e^{-i\delta\omega\tau}. \tag{I.B.70}$$

The first term returns to the real axis when $t = 2\tau$ independent of $\delta\omega$, but the second term evolves as if the second pulse had not occurred; it does not refocus.

Consider different values for θ:

$$\left.\begin{array}{l}\theta = 0, (1 - \cos \theta) = 0, (1 + \cos \theta) = 2\text{: no echo;}\\ \theta = \pi, (1 - \cos \theta) = 2, (1 + \cos \theta) = 0\text{: full echo} = M_0;\end{array}\right\} \tag{I.B.71}$$

and

$$\theta = \pi/2, (1 - \cos \theta) = 1, (1 + \cos \theta) = 1\text{: half echo} = M_0/2. \tag{I.B.72}$$

In a $\pi/2 - \tau - \pi/2$ sequence, the second $\pi/2$ has the same effect as applying 0 to half the spins and π to the rest.

Now try three pulses in the sequence: $\pi/2 - \tau_1 - \pi_2/2 - \tau_2 - \pi/2 - ?$ Consider nuclei with frequency $\omega_0 + \delta\omega$:

$$\delta m_0/\delta t = -i\,\delta\omega\, m_0, \tag{I.B.73}$$

solved (in the complex plane) by:

$$m_0 = A\, e^{-i\delta\omega t}. \tag{I.B.74}$$

We also consider the z-component of the magnetization. Before the first pulse, we have

$$m_0 = 0, \quad M_{\parallel} = M_0. \tag{I.B.75}$$

After the first $\pi/2$ pulse along the negative imaginary axis, this becomes

$$m_0 = M_0, \quad M_{\parallel} = 0. \tag{I.B.76}$$

Through the first period τ_1,

$$m_0 = M_0 e^{-i\delta\omega t}, \quad M_{\parallel} = 0. \tag{I.B.77}$$

Just before the second pulse, this is

$$m_0 = M_0 e^{-i\delta\omega\tau_1}, \quad M_{\parallel} = 0, \tag{I.B.78}$$

so that

$$m_0 = M_0(\cos \delta\omega\tau_1 - i \sin \delta\omega\tau_1). \tag{I.B.79}$$

The second $\pi/2$ pulse, along the positive real axis, produces

$$m_0 = M_0 \cos \delta\omega\tau_1 \tag{I.B.80}$$

and

$$M_{\parallel} = M_0 \sin \delta\omega\tau_1. \tag{I.B.81}$$

Through the second period, this evolves to

$$m_0 = M_0 \cos(\delta\omega\tau_1)e^{-i\delta\omega(t - \tau_1)} \tag{I.B.82}$$

and

$$M_{\parallel} = M_0 \sin \delta\omega\tau_1. \tag{I.B.83}$$

Similar considerations with respect to the third pulse lead to

$$m_0 = \frac{M_0}{4} \{\exp(i\delta\omega[2\tau_2 - t]) + \exp(i\delta\omega[2\tau_1 - t])$$

$$+ \exp(i\delta\omega[2\tau_2 - 2\tau_1 - t]) + \exp(i\delta\omega[-t])$$

$$+ 2 \exp(i\delta\omega[\tau_1 + \tau_2 - t]) - 2 \exp(i\delta\omega[\tau_2 - \tau_1 - t])\}. \tag{I.B.84}$$

The next-to-the-last term indicates that for

$$t = \tau_1 + \tau_2, \tag{I.B.85}$$

the term becomes independent of $\delta\omega$, and hence there will be a simulated echo, so called because the magnetization is "parked" along the z-axis between the last two pulses (see Fig. I.B.5). The other terms produce normal echoes whenever a bracketed term becomes zero. This is made clearer in the paper by Woessner referenced at the end of this chapter.

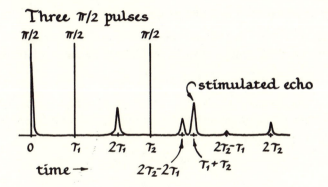

Figure I.B.5   Five echoes from three $\pi/2$ pulses.

I.B.6. T_2^* and "Hole Burning" – Homogeneous and Inhomogeneous Line Broadening

The Bloch Equation predicts through Eq. I.B.31 that the width of a line in continuous wave NMR at 1/2 maximum is inversely proportional to T_2, the spin–spin relaxation time. In fact, the experimental linewidth line may exceed this "natural" linewidth if the external magnetic field is sufficiently inhomogeneous (it is clear that the greater the spread in the external field, the greater the spread in resonance frequencies).

This distinction is also important in pulse NMR. Following a $\pi/2$ pulse the induction signal due to the magnetization precessing in the x–y plane will decay at a time much shorter than T_2; we will designate as T_2^* the actual decay time as influenced by magnetic field inhomogeneity. The π pulse in the spin echo experiment refocuses the spins that have lost phase coherence due to magnetic field inhomogeneity and will produce an echo. On the other hand, as we pointed out previously, the height of the echo is somewhat less than the height of the original FID; this reduction is due to the spin–spin relaxation.

This distinction between T_2 and T_2^* is also related to "hole burning". If one saturates the sample at a particular frequency in an inhomogeneously broadened line, then when a spectrum is obtained, that part of the line represented by the saturated frequency will be significantly reduced; a "hole is burned" in the line. If, however, the field is highly homogeneous and the linewidth is due solely to homogeneous broadening (spin–spin relaxation), you cannot "burn a hole in the line".

To summarize, T_2^* is a measure of phase coherence and is typically related to magnetic field inhomogeneity. Specifically, it is the time required for M_0 to drop to $1/e$ of its original value. The spin–spin relaxation time, T_2, is a phase memory time related to interactions among the spins themselves. In actual practice, the T_2^*-process is seldom first order so that T_2^* only approximates a first-order time constant. In liquids, the T_2 process *is* usually exponential.

I.B.7. The Bloch Equation and Diffusion

We will next see how the Bloch Equation handles the question of diffusion of magnetization. In one dimension, the diffusion equation has the form

$$\partial c/\partial t = D\partial^2 c/\partial x^2,$$ (I.B.86)

where c is the concentration of the diffusing species and D is the diffusion constant. This is sometimes called Fick's second law of diffusion.

In three dimensions, we have

$$\partial c/\partial t = D\nabla^2 c,$$ (I.B.87)

where

$$\nabla = \mathbf{i}\frac{\partial}{\partial x} + \mathbf{j}\frac{\partial}{\partial y} + \mathbf{k}\frac{\partial}{\partial z}$$ (I.B.88)

and

$$\nabla^2 = \nabla\cdot\nabla = \frac{\partial^2}{\partial x^2} + \frac{\partial^2}{\partial y^2} + \frac{\partial^2}{\partial z^2}.$$ (I.B.89)

We will use this as a guide to modifying the Bloch Equation to make it applicable to the diffusion of magnetization. With $c \to m$, we have

$$\partial m/\partial t = -i\omega_0 m - m/T_2 + D\nabla^2 m.$$ (I.B.90)

For diffusion to be detectable we need a concentration gradient; this is possible by introducing a gradient in the external magnetic field:

$$\mathbf{B} = \mathbf{B_0} + \mathbf{k}(\mathbf{r}\cdot\mathbf{G}).$$ (I.B.91)

$\mathbf{G}$ can have any direction but the field is still along $\mathbf{k}$.

Our modified Bloch Equation now becomes

$$\partial m/\partial t = -i[\omega_0 + \gamma\mathbf{r}\cdot\mathbf{G}]m - m/T_2 + D\nabla^2 m.$$ (I.B.92)

Without $\mathbf{G}$ and D we can write the solution as

$$m = \psi \exp\{-[i\omega_0 + (1/T_2)]t\}$$ (I.B.93)

where ψ is constant (see Eq. I.B.48). Now let ψ be a function of $\mathbf{r}$ and t, and use Eqs. I.B.92 and I.B.93.

$$\partial\psi/\partial t = -i\gamma(\mathbf{r}\cdot\mathbf{G})\psi + D\nabla^2\psi.$$ (I.B.94)

First, let us neglect the diffusion term and obtain

$$\partial\psi/\partial t = -i\gamma(\mathbf{r}\cdot\mathbf{G})\psi.$$ (I.B.95)

What happens if we do a spin echo experiment:

$$(\pi/2 - \tau - \pi - \tau - \text{echo})?$$ (I.B.96)

Following the $\pi/2$ pulse, until the π pulse ($0 < t < \tau$), we can easily show that the solution to Eq. I.B.95 is

$$\psi = A \exp[-i\gamma(\mathbf{r}\cdot\mathbf{G})t].$$ (I.B.97)

Similarly, following the π pulse $(t > \tau)$,

$$\psi = A \exp[-i\gamma(\mathbf{r} \cdot \mathbf{G})(t - 2\tau) + i\varphi]. \tag{I.B.98}$$

We set the arbitrary constant $\varphi = 0$, and obtain

$$\psi = A \exp\{-i\gamma(\mathbf{r} \cdot \mathbf{G})[t + (\xi - 1)\tau]\} \tag{I.B.99}$$

where A is a constant, and

$$\left.\begin{aligned} \xi &= +1 \qquad \text{for } 0 < t < \tau \\ \xi &= -1 \qquad \text{for } t > \tau. \end{aligned}\right\} \tag{I.B.100}$$

Note the echo at $t = 2\tau$, as expected.

We now go back to Eq. I.B.92 which includes D, and substitute Eq. I.B.99 for ψ, but allowing A to be a function of time. We obtain

$$dA/dt = -\gamma^2 D[Gt + (\xi - 1)G\tau]^2 A. \tag{I.B.101}$$

We integrate this differential equation to find

$$\ln[A(\tau')/A(0)] = -\gamma^2 DG^2 \left[\int_0^{\tau'} \{t^2 + 2(\xi - 1)t\tau + (\xi - 1)^2 \tau^2\} \, dt \right]$$
$$= -\gamma^2 DG^2 \left[\frac{\tau'^3}{3} - 2\tau\tau'^2 + A\tau^2\tau' - 2\tau^3 \right]. \tag{I.B.102}$$

At $\tau' = 2\tau$ (at the peak of the echo), this becomes

$$\ln[A(2\tau)/A(0)] = -\tfrac{2}{3}\gamma^2 DG^2 \tau^3. \tag{I.B.103}$$

This means that the amplitude at $t = 2\tau$ (when the echo occurs) is *diminished* by diffusion (see Fig. I.B.6).

As mentioned in Section I.B.5, we can use a number of successive Hahn echo experiments in which τ is varied to measure T_2. In the presence of diffusion, this experiment is complicated by the diminution of the signal just discussed. Moreover, the longer τ, the larger τ^3, and the greater the reduction of the signal amplitude due to diffusion. A technique due to Carr and Purcell helps to correct this: instead of a single Hahn sequence, one uses, after the

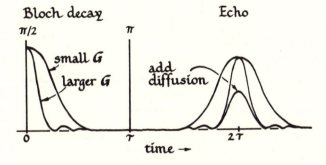

Figure I.B.6 Effect of diffusion on the spin echo.

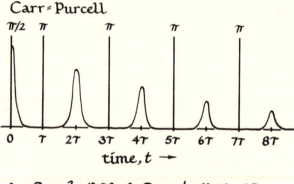

$$\ln R = -\tfrac{2}{3}\gamma^2 G^2 \tau^3 n D = -\tfrac{1}{3}\gamma^2 G^2 \tau^2 D t$$

Figure I.B.7   The Carr-Purcell sequence.

initial $\pi/2$ pulse, a series of π pulses, at τ, 3τ, 5τ, etc., which produces a set of echoes at 2τ, 4τ, 6τ, etc. (see Fig. I.B.7). The echo magnitudes decrease approximately exponentially with time constant T_2. The effect of diffusion is minimized by keeping short the interval between pulses so that τ^3 stays small.

A refinement of this technique involves the use of a time-dependent magnetic field gradient rather than the constant field described earlier. If the gradient is diminished during the rf pulses, the strength of the pulse does not have to be very large, and if one keeps the gradient small when the echo occurs, that echo will be wide and measuring its amplitude easy. When diffusion coefficients are measured using the spin echo method, the advantages of using a time-dependent gradient makes the technique more widely applicable. In addition, if the gradient pulses are short compared to their separation, the method is useful when one needs to determine precisely the time during which diffusion is being measured.

Experimentally, one varies the field gradient from a lower value g_0 to a higher value g and back with a pulse length δ for g and a period Δ (see Fig. I.B.8).

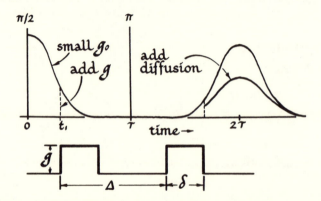

Figure I.B.8   The pulsed-field-gradient, spin-echo (PFGSE) diffusion measurement.

It can be shown that the echo amplitude satisfies

$$\ln[A/A_0] = -\tfrac{2}{3}\gamma^2 g_0^2 \tau^3 D$$
$$- \gamma^2 g^2 \delta^2 (\Delta - \tfrac{1}{3}\delta)D$$
$$+ \gamma^2 \mathbf{g}_0 \cdot \mathbf{g}\delta[(t_1^2 + t_2^2) + \delta(t_1 + t_2) + \tfrac{2}{3}\delta^2 - 2\tau^2]D, \quad \text{(I.B.104)}$$

where t_1 is the time the first gradient pulse starts and t_2 measures the time between the end of the second gradient pulse and the maximum of the echo. If $\mathbf{g}$ and $\mathbf{g}_0$ are chosen to be orthogonal, the final term can be made to vanish.

I.B.8. Chemical Exchange

A broad area in which NMR experiments can provide important information is *chemical exchange*, in which the nuclei being observed are moving between two magnetically inequivalent sites. In a very simple example, a nucleus may be jumping back and forth between two sites with different resonance frequencies f_A and f_X with an average lifetime τ in each site. It can be shown in a straightforward fashion using the Bloch equation in its complex form that there are two extreme experimental cases depending on the relative magnitude of $|f_A - f_B|$ and τ^{-1}. In the fast exchange limit ($|f_A - f_B| \ll \tau^{-1}$), one will observe a single intermediate line at frequency $(|f_A - f_B|)/2$. In the slow exchange limit ($|f_A + f_B| \gg \tau^{-1}$), one observes two separate lines at the two frequencies f_A and f_B.

The Bloch equation can be used to establish these results and to predict the experimental result for the case in which $|f_A - f_B|$ and τ^{-1} are of comparable magnitude. Similar analyses are possible for a number of time-dependent phenomena such as hindered rotation of molecules.

A more detailed treatment of chemical exchange can be found in the reference by Kaplan and Fraenkel [1980], listed at the end of this chapter.

I.C. NUCLEAR SPIN QUANTUM MECHANICS AND THE DENSITY MATRIX

In this section, we will review briefly some fundamentals of quantum mechanics and introduce the density matrix. Density matrix theory is well suited to the treatment of many NMR situations, and is widely used in the field, particularly in describing the effects of pulses on a spin system, and to treat nuclear magnetic relaxation.

I.C.1. Basic Quantum Mechanics

Classically, the Hamiltonian function H is defined in terms of generalized coordinates q_i and conjugate momenta p_i; in the cases that we will describe, H is identifiable with the energy E of the system described by H. We urge those readers unfamiliar with this formulation of classical mechanics to refer to the reference by Goldstein listed at the end of the chapter. The canonical transformation from classical mechanics to quantum mechanics is to make the substitutions

$$p_i \rightarrow (\hbar/i)\, \partial/\partial q_i \qquad\qquad (\text{I.C.1})$$

and

$$E \rightarrow i\hbar\, \partial/\partial t. \qquad\qquad (\text{I.C.2})$$

This leads to Schrödinger's Equation

$$H\Psi = i\hbar\, \partial\Psi/\partial t, \qquad\qquad (\text{I.C.3})$$

where Ψ, the state function, or wave function, of the system is a function of the coordinates and the time. One typically separates the partial differential equation by writing

$$\Psi = \psi \exp(-iEt/\hbar), \qquad\qquad (\text{I.C.4})$$

where ψ is a function of coordinates alone. Eq. I.C.3. becomes

$$H\psi = E\psi. \qquad\qquad (\text{I.C.5})$$

Eq. I.C.5 is an eigenvalue problem for the Hamiltonian operator H. One finds a set of eigenfunctions ϕ_n corresponding to energy eigenvalues E_n:

$$H\phi_n = E_n\phi_n. \qquad\qquad (\text{I.C.6})$$

The eigenfunctions satisfy orthonormality relations:

$$\int \phi_m^*\phi_n\, d\tau = (\phi_m, \phi_n) = \langle m|n \rangle = \delta_{mn}. \qquad\qquad (\text{I.C.7})$$

The operator H can now be written as a matrix with general element

$$H_{ij} = \int \phi_i^* H\phi_j\, d\tau = (\phi_i, H\phi_j)$$

$$= (\phi_i, E_j\phi_j) = E_j(\phi_i, \phi_j) = E_j\delta_{ij}. \qquad\qquad (\text{I.C.8})$$

In general, a wave function can be written as a linear combination of energy eigenfunctions:

$$\psi = \sum_i c_i\phi_i. \qquad\qquad (\text{I.C.9})$$

If the matrices representing two operators commute (see Appendix A1), it can be shown that they have closely related eigenfunctions (identical if there is no "degeneracy"; that is, if there is only one eigenfunction per eigenvalue).

One defines the expectation value of an operator F, which depends on the coordinates and momenta of the system, by

$$\langle F \rangle = \int \psi^* F\psi\, d\tau$$

$$= \sum_{i,j} c_i^* c_j(\phi_i, F\phi_j) = \sum_{i,j} c_i^* c_j\langle i|F|j \rangle, \qquad\qquad (\text{I.C.10})$$

where we have used Eq. I.C.9.

We can write the c_i and the ϕ_i as many-dimensional column vectors $\mathbf{c}$ and ϕ (see Appendix A1) so that

$$\psi = \tilde{\phi} \cdot \mathbf{c} = (\phi_1 \phi_2 \cdots) \begin{pmatrix} c_1 \\ c_2 \\ \vdots \end{pmatrix} = \tilde{\mathbf{c}} \cdot \phi = (c_1 c_2 \cdots) \begin{pmatrix} \phi_1 \\ \phi_2 \\ \vdots \end{pmatrix}. \quad \text{(I.C.11)}$$

Moreover, the expectation value of F can be written

$$\langle F \rangle = \mathbf{c}^\dagger F \mathbf{c} = (c_1^* c_2^* \cdots) \begin{pmatrix} F_{11} & F_{12} & \cdots \\ F_{21} & F_{22} & \cdots \\ \vdots & \vdots & \end{pmatrix} \begin{pmatrix} c_1 \\ c_2 \\ \vdots \end{pmatrix} \quad \text{(I.C.12)}$$

where

$$F_{ij} = \langle i|F|j \rangle. \quad \text{(I.C.13)}$$

I.C.2. The Angular Momentum

We now consider the dynamical variable which, along with H, is of most concern to us in NMR: the angular momentum $\mathbf{L}$. Classically, we have

$$\mathbf{L} = \mathbf{r} \times \mathbf{p}. \quad \text{(I.C.14)}$$

Using the quantum mechanical prescription given earlier (see Eq. I.C.1), we have

$$\mathbf{L} = \mathbf{r} \times \frac{\hbar}{i} \mathbf{V}. \quad \text{(I.C.15)}$$

By letting each component of $\mathbf{L}$ operate on an arbitrary function ψ, we can show

$$\left. \begin{aligned} [L_x, L_y] &= i\hbar L_z \\ [L_y, L_z] &= i\hbar L_x. \\ [L_z, L_x] &= i\hbar L_y \end{aligned} \right\} \quad \text{(I.C.16)}$$

A shorthand way to write this is

$$\mathbf{L} \times \mathbf{L} = i\hbar \mathbf{L}. \quad \text{(I.C.17)}$$

This also applies to spin angular momentum even though there is no momentum $\mathbf{p}$, as in Eq. I.C.14, to identify with spin:

$$\mathbf{S} \times \mathbf{S} = i\hbar \mathbf{S}. \quad \text{(I.C.18)}$$

For a spin 1/2 particle, we can write

$$\mathbf{S} = \frac{\hbar}{2} \sigma, \quad \text{(I.C.19)}$$

where σ has the Pauli spin matrices as components (see Eq. A.1.47 in Appendix A1). One easily shows that Eq. I.C.19 satisfies Eq. I.C.18 in matrix form.

It is standard to write the nuclear spin operator as

$$\mathbf{I}\hbar = \frac{\hbar}{2}\,\boldsymbol{\sigma}, \tag{I.C.20}$$

so that the operator $\mathbf{I}$ is dimensionless.

Associated with the nuclear spin is a magnetic moment μ (see Chapter I):

$$\boldsymbol{\mu} = \gamma\hbar\mathbf{I}, \tag{I.C.21}$$

which defines the magnetogyric ratio γ. In a magnetic field $\mathbf{B}_0$ in the z-direction, the energy of the dipole is

$$H = -\boldsymbol{\mu}\cdot\mathbf{B}_0 = -\gamma\hbar B_0 I_z. \tag{I.C.22}$$

Clearly, the relevant commutators satisfy

$$[H, I_z] = 0; \; \lfloor I_z, I^2 \rfloor = \lfloor I_z, (I_x^2 + I_y^2 + I_z^2) \rfloor = 0; \; \lfloor H, I^2 \rfloor = 0. \tag{I.C.23}$$

Since I_z, I^2 and H all commute with one another, they have simultaneous eigenfunctions (see previous discussion). Reference to Eqs. A.1.46 and A.1.47 indicates that $v(1/2) = \alpha$ is an eigenfunction of I_z with an eigenvalue 1/2; that is,

$$I_z\alpha = \frac{1}{2}\begin{pmatrix} 1 & 0 \\ 0 & -1 \end{pmatrix}\begin{pmatrix} 1 \\ 0 \end{pmatrix} = \frac{1}{2}\begin{pmatrix} 1 \\ 0 \end{pmatrix} = \tfrac{1}{2}\alpha. \tag{I.C.24}$$

α is also an eigenfunction of I^2 with eigenvalue $\tfrac{3}{4}$, and of H with eigenvalue $-\tfrac{1}{2}\gamma\hbar B_0$. α represents a "spin up" state. The results for β follow in a similar way.

I.C.3. Perturbation Theory

In many cases, the Hamiltonian of a system can be divided into two pieces

$$H = H_0 + H_1, \tag{I.C.25}$$

where H_1 is in some sense small in comparison with H_0, and where the solution to Schrödinger's equation is known if $H_1 = 0$. In such cases, a powerful formalism, called *perturbation theory*, is available to approximate the solution to Schrödinger's equation for the full Hamiltonian given in Eq. I.C.25. The fundamental results depend on whether or not H_1 is a function of time.

If H_1 is *not* dependent on time, its effect is to cause a small shift in the energy levels for the unperturbed ($H_1 = 0$) case. The shift in the nth level is

$$\Delta E_n = \int \phi_n^* H_1 \phi_n \, d\tau = \langle n|H_1|n\rangle, \tag{I.C.26}$$

where ϕ_n is the nth unperturbed eigenfunction; that is, the shift is equal to the expectation value of the perturbation with respect to the nth unperturbed state.

On the other hand, if H_1 *is* a function of time, its effect is to induce transitions between the states of the unperturbed Hamiltonian H_0. If ϕ_m and ϕ_n are two different eigenfunctions of H_0, the transition probability due to H_1 between states n and m is

$$W_{ij} = (1/\hbar^2 t) \left| \int_0^t \langle m_j | H'(t) | m_i \rangle \exp(-i\omega_{ij}t)\, dt' \right|^2 \qquad \text{(I.C.27)}$$

where

$$\omega_{ij} = (E_j - E_i)/\hbar. \qquad \text{(I.C.28)}$$

I.C.4. The Density Matrix and the Heisenberg Equation of Motion

The density matrix form of quantum mechanics is well suited to the treatment of many NMR situations, and is widely used in the field, particularly in describing the effects of pulses on a spin system, and to treat nuclear magnetic relaxation.

We wish to develop means to evaluate the expectation value $\langle F \rangle$ and also its time dependence. Eqs. I.C.9 and I.C.10 lead us into a discussion of the density matrix. We can write

$$\langle F \rangle = \sum_{i,j} c_i^* c_j F_{ij} = \sum_j \sum_i P_{ji} F_{ij} \qquad \text{(I.C.29)}$$

where we have defined

$$P_{ji} = c_i^* c_j, \qquad \text{(I.C.30)}$$

or, in vector form,

$$\mathbf{P} = \mathbf{c}\mathbf{c}^\dagger. \qquad \text{(I.C.31)}$$

But

$$\sum_i P_{ji} F_{ij} = (PF)_{jj}, \qquad \text{(I.C.32)}$$

(see Eq. A.1.16), so

$$\langle F \rangle = \sum_j (PF)_{jj} = \text{Tr}(PF), \qquad \text{(I.C.33)}$$

where we have used Eqs. I.C.29 and I.C.32.

For an ensemble of systems described by Ψ, we can define a density matrix ρ by

$$\rho = \sum_l p_l P_l, \qquad \text{(I.C.34)}$$

a weighted average over all the states in the ensemble (p_l is the normalized statistical weighting factor for the lth member of the ensemble).

Using Eqs. I.C.33 and I.C.34, we have for the ensemble average

$$\langle \bar{F} \rangle = \text{Tr}(\rho F) = \text{Tr}(F\rho). \qquad \text{(I.C.35)}$$

Recall that the ϕ_i and F are time-independent. The c_i may be time-dependent.

One way we can determine the time-dependence of F is as follows:

$$\frac{d}{dt}\langle F \rangle = \frac{d}{dt}\int \Psi^* F \Psi \, d\tau$$

$$= \int \frac{\partial \Psi^*}{\partial t} F \Psi \, d\tau + \int \Psi^* \frac{\partial F}{\partial \tau} \Psi \, d\tau + \int \Psi^* F \frac{\partial \Psi}{\partial t} \, d\tau. \quad \text{(I.C.36)}$$

By using Eq. I.C.3 and its complex conjugate

$$(H\Psi)^* = -i\hbar \frac{\partial \Psi^*}{\partial \tau}, \quad \text{(I.C.37)}$$

we obtain

$$\frac{d}{dt}\int \Psi^* F \Psi \, d\tau = \int (-i\hbar)(H\Psi)^* F \Psi \, d\tau + \int \Psi^* \frac{\partial F}{\partial t} \Psi \, d\tau + \int (i\hbar)\Psi^* F(H\Psi) \, d\tau.$$

$$\text{(I.C.38)}$$

Now H is a Hermitian operator (see Eq. A1.24); all operators representing dynamical variables are Hermitian since these have *real* eigenvalues, and hence represent measurable quantities.

From Eq. A1.24, then, we have

$$H_{ij}^\dagger = H_{ji}^* = H_{ij}. \quad \text{(I.C.39)}$$

This enables us to rewrite the first term on the right-hand-side of Eq. I.C.38; combining terms, we have

$$\frac{d}{dt}\int \psi^* F \psi \, d\tau = \frac{i}{\hbar}\int \Psi^*(HF - FH)\Psi \, d\tau + \int \Psi^* \frac{\partial F}{\partial t} \Psi \, d\tau. \quad \text{(I.C.40)}$$

Since this holds for arbitrary Ψ, we have

$$\frac{dF}{dt} = \frac{\partial F}{\partial t} + \frac{i}{\hbar}[H, F], \quad \text{(I.C.41)}$$

which is called the Heisenberg equation of motion.

Consider the time dependence of the z-component of the spin:

$$\frac{dI_z}{dt} = \frac{i}{\hbar}[B, I_z] = 0$$

$$\frac{dI_x}{dt} = \frac{i}{\hbar}[B, I_x] = \frac{i}{\hbar}(-\gamma\hbar B_0)[I_z, I_x]$$

$$= -i\gamma B_0(iI_y) = +\gamma B_0 I_y. \quad \text{(I.C.42)}$$

Similarly

$$\frac{dI_y}{dt} = -\gamma B_0 I_x. \quad \text{(I.C.43)}$$

The vector representation, then, is

$$d\mathbf{I}/dt = \gamma \begin{pmatrix} \hat{i} & \hat{j} & \hat{p} \\ I_x & I_y & I_z \\ 0 & 0 & B_0 \end{pmatrix} = +\gamma B_0 I_y \hat{i} - \gamma B_0 I_x \hat{j}, \qquad (\text{I.C.44})$$

so that

$$d\mathbf{I}/dt = \mathbf{I} \times \gamma \mathbf{B_0} \quad \text{or} \quad (\boldsymbol{\omega}_0 \times \mathbf{I} \text{ if } -\boldsymbol{\omega}_0 = \gamma \mathbf{B_0}) \qquad (\text{I.C.45})$$

or

$$\frac{d\langle \boldsymbol{\mu} \rangle}{dt} = \langle \boldsymbol{\mu} \rangle \times \gamma \mathbf{B_0}. \qquad (\text{I.C.46})$$

This mathematical curiosity may help explain why the classical vector picture of precession works so well.

There is another way to obtain the time dependence of $\langle F \rangle$. If we represent an ensemble average by a line over the quantities involved, then, from Eq. I.C.34, we have for the general matrix element of the density matrix

$$\rho_{nm} = \overline{c_m^* c_n} = \langle n|\rho|m \rangle. \qquad (\text{I.C.47})$$

We can expand the time-dependent wave function Ψ as a linear combination of time-independent eigenfunctions,

$$\Psi = \sum_k c_k \phi_k, \qquad (\text{I.C.48})$$

if we allow the c_k to be time-dependent. From Eqs. I.C.3 and I.C.9, we have

$$-\frac{\hbar}{i} \sum_k \frac{dc_k}{dt} \phi_k = \sum_k c_k H \phi_k. \qquad (\text{I.C.49})$$

If we multiply on the left by ϕ_n^* and integrate, Eq. I.C.49 becomes

$$-\frac{\hbar}{i} \sum_k \frac{dc_k}{dt} \langle n|k \rangle = -\frac{\hbar}{i} \frac{dc_n}{dt} = \sum_k c_k \langle n|H|k \rangle \sum_k c_k \langle n|H|k \rangle, \qquad (\text{I.C.50})$$

since $\langle n|k \rangle = \int \phi_n^* \phi_k \, d\tau = \delta_{nk}$ from the orthonormality of the eigenfunctions.

Recall from Eq. I.C.30

$$\langle n|P|m \rangle = c_n c_m^*, \qquad (\text{I.C.51})$$

and differentiate with respect to time:

$$\frac{d}{dt} \langle n|P|m \rangle = \frac{d}{dt} (c_n c_m^*) = c_n \frac{dc_m^*}{dt} + \frac{dc_n}{dt} c_m^*$$

$$= \frac{i}{\hbar} \sum_k \{ c_n c_k^* \langle k|H|m \rangle - \langle n|H|k \rangle c_k c_m^* \}, \qquad (\text{I.C.52})$$

where we have used Eqs. I.C.50 and I.C.51. We now have

$$\frac{dP_{nm}}{dt} = \frac{i}{\hbar} \sum_k \{ \langle n|P|k \rangle \langle k|H|m \rangle - \langle n|H|k \rangle \langle k|P|m \rangle \}$$

$$= \frac{i}{\hbar} \{ \langle n|PH|m \rangle - \langle n|HP|m \rangle \} = \frac{i}{\hbar} \langle n|[P, H]|m \rangle, \qquad (\text{I.C.53})$$

where we have used Eq. I.C.7 to show

$$\sum_k |k\rangle\langle k| = 1. \tag{I.C.54}$$

In operator form, then,

$$\frac{dP}{dt} = \frac{i}{\hbar}[P, H]. \tag{I.C.55}$$

Do not confuse this relation with the Heisenberg equation of motion, which it so closely resembles.

Taking an ensemble average, we arrive at the equation of motion for the density matrix:

$$\frac{d\rho}{dt} = \frac{i}{\hbar}[\rho, H]. \tag{I.C.56}$$

It can be shown by substitution that the solution of Eq. I.C.56 is

$$\rho(t) = e^{-(i/\hbar)Ht}\rho(0)e^{+(i/\hbar)Ht} = U_H(t)\rho(0)U_H^{-1}(t) \tag{I.C.57}$$

if H is time-dependent; we have written $U_H(t) = e^{-(i/\hbar)Ht}$.

If we are given $\rho(0)$ and H, we can calculate $\rho(t)$, and then using Eq. I.C.35, we can find the ensemble average of the expectation value of any dynamical variable F as a function of time.

I.C.5. NMR and the Density Matrix

In this section and the two following, we draw heavily on the excellent presentation of the subject appearing in Farrar and Harriman [1989], given at the end of this chapter.

Consider the simple case of a spin 1/2 nucleus in a magnetic field $\mathbf{B}_0$. From Eq. I.C.22, we have

$$H = -\hbar_\gamma B_0 I_z = -\hbar\omega I_z, \qquad \omega = \gamma B_0. \tag{I.C.58}$$

From Eqs. I.C.58 and A1.47,

$$H = -\hbar\omega I_z = \begin{pmatrix} -\dfrac{\hbar\omega}{2} & 0 \\ 0 & \dfrac{\hbar w}{2} \end{pmatrix}, \tag{I.C.59}$$

so that

$$U_H = \begin{pmatrix} e^{i\omega t/2} & 0 \\ 0 & e^{-i\omega t/2} \end{pmatrix}. \tag{I.C.60}$$

U_H is an example of a *unitary* transformation, one for which

$$U^{-1} = U^\dagger; \tag{I.C.61}$$

that is, one for which the inverse equals the Hermitian adjoint. Equation

I.C.61 is an example of a general type of transformation

$$A' = UAU^{-1}. \tag{I.C.62}$$

It can be shown that such transformations indicated by Eqs. I.C.61 and I.C.62 are equivalent in a generalized sense to a rotation of coordinate axes that leaves vector lengths unchanged. See also Eqs. A1.25 through A1.37.

We next approach the question of finding the thermal equilibrium density matrix. Since the ϕ_n are normalized, the probability that the system described by the wave function Ψ for one member of the ensemble is

$$|c_m|^2 = c_m c_m^*. \tag{I.C.63}$$

When averaged over the ensemble, we obtain the Boltzmann equilibrium distribution,

$$\overline{c_m c_m^*} = \frac{e^{-E_m/kT}}{Z}, \tag{I.C.64}$$

where Z is the normalizing "sum over states":

$$Z = \sum_m e^{-E_m/kT}. \tag{I.C.65}$$

In thermal equilibrium, the "random phases" assumption leads to

$$\overline{c_n c_m^*} = 0 \quad \text{if } m \neq n. \tag{I.C.66}$$

We have, then,

$$\langle n|\rho|m \rangle = (\delta_{nm}/Z)e^{-E_n/kT}. \tag{I.C.67}$$

For a single spin $1/2$ particle, this gives

$$\rho_{eq} = \frac{1}{Z}\begin{pmatrix} \exp(\hbar\omega/2kT) & 0 \\ 0 & \exp(-\hbar\omega/2kT) \end{pmatrix}, \tag{I.C.68}$$

where we have used Eqs. I.C.64 and I.C.65. At room temperature,

$$\hbar\omega \ll kT, \tag{I.C.69}$$

so

$$e^{\hbar\omega/2kT} \approx 1 + \hbar\omega/2kT, \qquad e^{-\hbar\omega/2kT} \approx 1 - \hbar\omega/2kT \tag{I.C.70}$$

and

$$Z \approx 2, \tag{I.C.71}$$

so

$$\rho_{eq} = \frac{1}{2}\begin{pmatrix} 1 + \dfrac{\hbar\omega}{2kT} & 0 \\ 0 & 1 - \dfrac{\hbar\omega}{2kT} \end{pmatrix}, \tag{I.C.72}$$

If we define

$$\frac{\hbar\omega}{2kT} = p, \tag{I.C.73}$$

Eq. I.C.72 becomes

$$\rho_{eq} = \frac{1}{2}\begin{pmatrix} 1 & 0 \\ 0 & 1 \end{pmatrix} + \frac{1}{2}\begin{pmatrix} p & 0 \\ 0 & -p \end{pmatrix} = \tfrac{1}{2}\sigma_0 + \frac{p}{2}\sigma_z, \qquad (I.C.74)$$

where we have used Eqs. I.C.73 and A1.47. Since σ_0 is just the two-by-two unit matrix, it will commute with any U so that, according to Eq. I.C.57, it will not evolve in time under any circumstances. We will drop it in future considerations.

We have, then

$$\rho_{eq} = \frac{p}{2}\sigma_z = \frac{p}{2}\begin{pmatrix} 1 & 0 \\ 0 & -1 \end{pmatrix} = pI_z. \qquad (I.C.75)$$

I.C.6. Pulses and the Density Matrix

Another type of unitary transformation of importance in pulse NMR is one bringing about a rotation of the nuclear magnetization through the use of a pulse. The transformation corresponding to rotating the magnetization through an angle φ about the x-axis is

$$U_{\phi x} = \begin{pmatrix} \cos(\varphi/2) & -i\sin(\varphi/2) \\ -i\sin(\varphi/2) & \cos(\varphi/2) \end{pmatrix}. \qquad (I.C.76)$$

The sign of φ is called positive if it results in a clockwise rotation looking from the origin out along the x-axis.

We will show that Eq. I.C.76 gives the right answer for a pulse that rotates the magnetization from the z-axis to the y-axis. In this case, we use $\varphi = -\pi/2$:

$$U_{-(\pi/2)x} = \begin{pmatrix} \cos(-\pi/4) & -i\sin(-\pi/4) \\ -i\sin(-\pi/4) & \cos(-\pi/4) \end{pmatrix} = 2^{-1/2}\begin{pmatrix} 1 & i \\ i & 1 \end{pmatrix} \qquad (I.C.77)$$

and

$$U_{-(\pi/2)x}^{-1} = U_{-(\pi/2)x}^{\dagger} = 2^{-1/2}\begin{pmatrix} 1 & -i \\ -i & 1 \end{pmatrix}. \qquad (I.C.78)$$

Therefore

$$\rho = U_{-(\pi/2)x}\rho_{eq}U_{-(\pi/2)x}^{-1}$$

$$= \frac{p}{4}\begin{pmatrix} 1 & i \\ i & 1 \end{pmatrix}\begin{pmatrix} 1 & 0 \\ 0 & -1 \end{pmatrix}\begin{pmatrix} 1 & -i \\ -i & 1 \end{pmatrix} = \frac{p}{2}\begin{pmatrix} 0 & -i \\ i & 0 \end{pmatrix} = \frac{p}{2}\sigma_y = pI_y. \qquad (I.C.79)$$

Using the new ρ, we calculate the new magnetization in the x–y plane, in terms of the complex operator $I_+ = Ix + iIy$ which represents the x- and y-components of magnetization as real and imaginary parts, respectively, see Eq. A1.52

$$\langle I_x + iI_y \rangle_{t=0} = \mathrm{Tr}\{(I_x + iI_y)(pI_y)\} = \mathrm{Tr}(ipI_y^2) = \frac{ip}{2}, \qquad (I.C.80)$$

and find it to be along the imaginary axis (the y-axis), and of magnitude equal to the equilibrium value along the z-axis, so Eq. I.C.76 works in describing the effect of a $-\pi/2$ rotation about the x-axis. A rotation about the y-axis is given by

$$U_{\varphi y} = \begin{pmatrix} \cos(\varphi/2) & -\sin(\varphi/2) \\ \sin(\varphi/2) & \cos(\varphi/2) \end{pmatrix}. \qquad \text{(I.C.81)}$$

We now use Eqs. I.C.79 and I.C.57 to see how the density matrix evolves after the $-(\pi/2)x$ pulse:

$$\rho(t) = U_H \rho(0) U_H^{-1} = \frac{p}{2} U_H \sigma_y U_H^{-1}$$

$$= \frac{p}{2} \begin{pmatrix} e^{i\omega t/2} & 0 \\ 0 & e^{-i\omega t/2} \end{pmatrix} \begin{pmatrix} 0 & -i \\ i & 0 \end{pmatrix} \begin{pmatrix} e^{-i\omega t/2} & 0 \\ 0 & e^{i\omega t/2} \end{pmatrix}$$

$$= \frac{p}{2} \begin{pmatrix} 0 & -i e^{i\omega t} \\ i e^{-i\omega t} & 0 \end{pmatrix}$$

$$= \frac{p}{2} \begin{pmatrix} 0 & -i \cos \omega t + \sin \omega t \\ i \cos \omega t + \sin \omega t & 0 \end{pmatrix}$$

$$= \frac{p}{2} \{(\cos \omega t)\sigma_y + (\sin \omega t)\sigma_x\}. \qquad \text{(I.C.82)}$$

The magnetization at time t in the x–y plane, then, is

$$= \langle I_x + i I_y \rangle_{t=t} = \text{Tr}\{(I_x + i I_y)\rho(t)\} = \frac{ip}{2} e^{-i\omega t}, \qquad \text{(I.C.83)}$$

and is seen to rotate with unchanged magnitude in the x–y plane with angular frequency ω, as expected.

It will be convenient to have an additional unitary operator which will enable us to transform to the rotating frame, as described in Chapter I. It can be shown that the appropriate matrix is

$$U_\Omega(t) = \begin{pmatrix} \exp(-i\Omega t/2) & 0 \\ 0 & \exp(i\Omega t/2) \end{pmatrix}, \qquad \text{(I.C.84)}$$

for a frame rotating with angular frequency Ω. We will demonstrate that this works for the example just given, a spin $1/2$ nuclear magnetization rotating in the x–y plane:

$$\rho'(t) = U_\Omega \rho(t) U_\Omega^{-1}$$

$$= \frac{p}{2} \begin{pmatrix} e^{-i\Omega t/2} & 0 \\ 0 & e^{+i\Omega t/2} \end{pmatrix} \begin{pmatrix} 0 & -i e^{i\omega t} \\ i e^{-i\omega t} & 0 \end{pmatrix} \begin{pmatrix} e^{i\Omega t/2} & 0 \\ 0 & e^{-i\Omega t/2} \end{pmatrix}$$

$$= \frac{p}{2} \begin{pmatrix} 0 & -i e^{-i(\Omega - \omega)t} \\ i e^{i(\Omega - \omega)t} & 0 \end{pmatrix}. \qquad \text{(I.C.85)}$$

Use of this in Eq. I.C.83 leads to

$$\langle I_x + i I_y \rangle = \frac{ip}{2} e^{(i\Omega - \omega)t}. \qquad \text{(I.C.86)}$$

If $\Omega = 0$ (frame not rotating), the answer reduces to Eq. I.C.83. If $\Omega = \omega$, the magnetization appears stationary in the rotating frame.

We now apply these ideas to the case of a spin echo sequence: a $-(\pi/2)x$ pulse, an evolution time, then a π_y pulse; then look for an echo after another evolution period. Just before the first pulse, we have, from Eq. I.C.75,

$$\rho'(a) = \frac{p}{2}\sigma_z. \tag{I.C.87}$$

Just after the $-(\pi/2)x$ pulse, we have, from Eq. I.C.78,

$$\rho'(b) = \frac{p}{2}\sigma_y. \tag{I.C.88}$$

After the evolution time s but before the π_y pulse, the density matrix becomes

$$\rho'(c) = \frac{p}{2}\begin{pmatrix} 0 & -ie^{-iqs} \\ ie^{iqs} & 0 \end{pmatrix}, \tag{I.C.89}$$

where we have written

$$q = \Omega - \omega \tag{I.C.90}$$

and have used Eq. I.C.85.

The form of the π_y pulse operator is (see Eq. I.C.81)

$$U_{\pi y} = \begin{pmatrix} 0 & -1 \\ 1 & 0 \end{pmatrix} = -U_{\pi y}^{-1}, \tag{I.C.91}$$

so after the pulse

$$\rho'(d) = \frac{p}{2}\begin{pmatrix} 0 & -1 \\ 1 & 0 \end{pmatrix}\begin{pmatrix} 0 & -ie^{-iqs} \\ ie^{iqs} & 0 \end{pmatrix}\begin{pmatrix} 0 & 1 \\ -1 & 0 \end{pmatrix}$$

$$= \frac{p}{2}\begin{pmatrix} 0 & -ie^{iqs} \\ ie^{-iqs} & 0 \end{pmatrix}. \tag{I.C.92}$$

Now let the system evolve a further time interval s; using Eq. I.C.57 we obtain

$$\rho'(e) = \frac{p}{2}\begin{pmatrix} e^{-iqs/2} & 0 \\ 0 & e^{iqs/2} \end{pmatrix}\begin{pmatrix} 0 & -ie^{-iqs} \\ ie^{-iqs} & 0 \end{pmatrix}\begin{pmatrix} e^{iqs/2} & 0 \\ 0 & e^{-iqs/2} \end{pmatrix}$$

$$= \frac{p}{2}\sigma_y = \rho'(b). \tag{I.C.93}$$

That is, the density matrix at e has returned to its value at b, and there is a *spin echo*.

In this derivation, we have used the form of U_H in the rotating frame:

$$U_H' = U_\Omega U_H = \begin{pmatrix} e^{-i\Omega t/2} & 0 \\ 0 & e^{+i\Omega t/2} \end{pmatrix}\begin{pmatrix} e^{i\omega t/2} & 0 \\ 0 & e^{-i\omega t/2} \end{pmatrix} = \begin{pmatrix} e^{-iqt/2} & 0 \\ 0 & e^{+iqt/2} \end{pmatrix}.$$

$$\tag{I.C.94}$$

I.C.7. The Two-Spin Case

How do we treat the case of more than one spin 1/2 nucleus? To begin with, recall that ρ for a single spin case has the form

$$
\left.\begin{array}{c}
\quad |\alpha\rangle \quad |\beta\rangle \\
\quad \downarrow \quad\; \downarrow \\
\rho = \begin{pmatrix} \rho_{11} & \rho_{12} \\ \rho_{21} & \rho_{22} \end{pmatrix} \quad \begin{array}{l} \leftarrow |\alpha\rangle \\ \leftarrow |\beta\rangle. \end{array}
\end{array}\right\}
\tag{I.C.95}
$$

A generalization of this to cover the two-spin case is

$$
\left.\begin{array}{c}
\rho = \begin{pmatrix} \rho_{11} & \rho_{12} & \rho_{13} & \rho_{14} \\ \rho_{21} & \rho_{22} & \rho_{23} & \rho_{24} \\ \rho_{31} & \rho_{32} & \rho_{33} & \rho_{34} \\ \rho_{41} & \rho_{42} & \rho_{43} & \rho_{44} \end{pmatrix} \begin{array}{l} \leftarrow |\alpha\alpha\rangle \\ \leftarrow |\alpha\beta\rangle \\ \leftarrow |\beta\alpha\rangle \\ \leftarrow |\beta\beta\rangle \end{array} \\
\qquad \uparrow \quad\;\; \uparrow \quad\;\; \uparrow \quad\;\; \uparrow \\
\quad |\alpha\alpha\rangle \; |\alpha\beta\rangle \; |\beta\alpha\rangle \; |\beta\beta\rangle
\end{array}\right\}
\tag{I.C.96}
$$

where

$$
|\alpha\alpha\rangle = |\alpha\rangle|\alpha\rangle, \quad \text{etc.}
\tag{I.C.97}
$$

The matrix shown in Eq. I.C.95 can always be constructed from a linear combination of the 3 (2×2) Pauli matrices and the identity matrix. In this case, we need 15 (4×4) matrices plus the identity matrix. Suppose A_x, A_y, and A_z are the matrices representing the components of the spin of nucleus one and B_x, B_y, and B_z the corresponding matrices for nucleus two. Combinations of these components and the identity matrix yield the 16 (4×4) matrices required.

We can calculate element by element a 4×4 matrix corresponding to a product of matrices as follows:

$$
\begin{aligned}
(A_z B_x)_{12} &= \langle \alpha | A_z | \alpha \rangle \langle \alpha | B_x | \beta \rangle \\
&= \langle \alpha | \tfrac{1}{2}\sigma_z | \alpha \rangle \langle \alpha | \tfrac{1}{2}\sigma_x | \beta \rangle \\
&= \frac{1}{4}\left[(1 \;\; 0)\begin{pmatrix} 1 & 0 \\ 0 & -1 \end{pmatrix}\begin{pmatrix} 1 \\ 0 \end{pmatrix}\right]\left[(1 \;\; 0)\begin{pmatrix} 0 & 1 \\ 1 & 0 \end{pmatrix}\begin{pmatrix} 0 \\ 1 \end{pmatrix}\right], \\
&= \tfrac{1}{4}(1)(1) = \tfrac{1}{4},
\end{aligned}
\tag{I.C.98}
$$

and so on for the fifteen other elements.

There is an easier way, however; we can define the direct product of two 2×2 matrices by the 4×4 matrix (see Eq. A1.39):

$$
M \otimes N = \begin{pmatrix} M_{11}N & M_{12}N \\ M_{21}N & M_{22}N \end{pmatrix}.
\tag{I.C.99}
$$

To use our earlier example,

$$A_z B_x = \tfrac{1}{2}\sigma_z \otimes \tfrac{1}{2}\sigma_x = \frac{1}{4}\begin{pmatrix} 1 & 0 \\ 0 & -1 \end{pmatrix} \otimes \begin{pmatrix} 0 & 1 \\ 1 & 0 \end{pmatrix}$$

$$= \frac{1}{4}\begin{pmatrix} 0 & 1 & 0 & 0 \\ 1 & 0 & 0 & 0 \\ 0 & 0 & 0 & -1 \\ 0 & 0 & -1 & 0 \end{pmatrix}. \tag{I.C.100}$$

Note that

$$(A_z B_x)_{12} = \tfrac{1}{4} \tag{I.C.101}$$

as we determined directly before.

The component of a single nucleus above is, of course, a 4 × 4 matrix to be calculated as follows:

$$A_y = \tfrac{1}{2}\sigma_y \otimes \sigma_0 = \frac{1}{2}\begin{pmatrix} 0 & -i \\ i & 0 \end{pmatrix} \otimes \begin{pmatrix} 1 & 0 \\ 0 & 1 \end{pmatrix}$$

$$= \frac{1}{2}\begin{pmatrix} 0 & 0 & -i & 0 \\ 0 & 0 & 0 & i \\ i & 0 & 0 & 0 \\ 0 & i & 0 & 0 \end{pmatrix}. \tag{I.C.102}$$

A random element can be checked. For example

$$(A_y)_{42} = \langle \beta | \tfrac{1}{2}\sigma_y | \alpha \rangle \langle \beta | \beta \rangle = \left[\tfrac{1}{2}(0 \quad 1)\begin{pmatrix} 0 & -i \\ i & 0 \end{pmatrix}\begin{pmatrix} 1 \\ 0 \end{pmatrix} \right] 1$$

$$= (\tfrac{1}{2}i)(1) = \tfrac{1}{2}i. \tag{I.C.103}$$

as we obtained above.

In an entirely similar fashion, other operators such as the Hamiltonian and rotation matrices can be written using the direct product in 4 × 4 matrix form for the two spin $\tfrac{1}{2}$ system. Eqs. I.C.57 and I.C.85 hold as they are with the understanding that matrix representations are now 4 × 4 instead of 2 × 2.

After reference to Eq. I.C.96, consider this representation of the density matrix:

$$\rho = \begin{pmatrix} POP & SQC & SQC & DQC \\ SQC & POP & ZQC & SQC \\ SQC & ZQC & POP & SQC \\ DQC & SQC & SQC & POP \end{pmatrix}. \tag{I.C.104}$$

The meanings of the symbols follow:

POP = populations of the energy states;

SQC = single quantum coherence; observable transitions;

ZQC = zero quantum coherence; neither spin changes orientation;

DQC = double quantum coherence; both spins change orientation.

We summarize important results in Table I.C.1 and Table I.C.2.

Table I.C.1   Important Quantities for a Two-Spin AX System

Pulses:

$$U_{-(\pi/2)x} = \frac{\sqrt{2}}{2}\begin{pmatrix} 1 & i \\ i & 1 \end{pmatrix} \quad U_{(\pi/2)y} = \frac{\sqrt{2}}{2}\begin{pmatrix} 1 & -1 \\ 1 & 1 \end{pmatrix} \quad U_{-\pi x} = \begin{pmatrix} 0 & i \\ i & 0 \end{pmatrix}$$

$$i = \sqrt{-1}$$

$$(U_{-(\pi/2)x})_A = U_{-(\pi/2)x} \otimes \sigma_0 \qquad (U_{(\pi/2)y})_A = U_{(\pi/2)y} \otimes \sigma_0 \qquad \sigma_0 = \begin{pmatrix} 1 & 0 \\ 0 & 1 \end{pmatrix}$$

$$(U_{-\pi x})_A = U_{-\pi x} \otimes \sigma_0 \qquad (U_{-\pi x})_x = \sigma_0 \otimes U_{-\pi x}$$

$$(U_{-(\pi/2)x})_x = \sigma_0 \otimes U_{-(\pi/2)x}.$$

Multiply for simultaneous pulses.

Spin Operators:

$$A_z = \tfrac{1}{2}\sigma_z \otimes \sigma_0 \qquad \sigma_z = \begin{pmatrix} 1 & 0 \\ 0 & -1 \end{pmatrix} = 2I_z$$

$$A_x = \tfrac{1}{2}\sigma_x \otimes \sigma_0 \qquad A_y = \tfrac{1}{2}\sigma_y \otimes \sigma_0 \qquad \sigma_x = \begin{pmatrix} 0 & 1 \\ 1 & 0 \end{pmatrix} = 2I_x$$

$$X_x = \tfrac{1}{2}\sigma_0 \otimes \sigma_x \qquad X_y = \tfrac{1}{2}\sigma_0 \otimes \sigma_y \qquad \sigma_y = \begin{pmatrix} 0 & -i \\ i & 0 \end{pmatrix} = 2I_y$$

$$A_+ = A_x + iA_y \qquad X_+ = X_x + iX_y \qquad X_z = \tfrac{1}{2}\sigma_0 \otimes \sigma_z$$

Expectation Values: $\langle \bar{F} \rangle = \mathrm{Tr}\,\rho F = \mathrm{Tr}\,F\rho$

$$\langle \bar{A}_z \rangle = \tfrac{1}{2}(\rho_{11} + \rho_{22} - \rho_{33} - \rho_{44})$$

$$\langle \bar{X}_z \rangle = \tfrac{1}{2}(\rho_{11} - \rho_{22} + \rho_{33} - \rho_{44})$$

$$\langle \bar{A}_+ \rangle = \rho_{31} + \rho_{42} \qquad \langle \bar{X}_+ \rangle = \rho_{21} + \rho_{43}$$

Table I.C.2   Qualitative Description of the Basis Matrices (Courtesy of J. P. Burgess)

A_x, A_y, A_z X_x, X_y, X_z	Matrix representations of the x, y, z, components of the A or X spins. These represent single quantum coherence between states that differ by one quanta, that is 1–2, 1–3 or 1–3, 2–4.
$A_x X_x, A_x X_y$ $A_y A_x, A_y X_y$	Combination zero and double quantum coherence. It is unobservable, but these states allow polarization transfer between A and X. These states are easily recognized by elements along the reverse diagonal. They link states that differ by zero quanta like 2–3, or two quanta like 1–4.
$A_x X_z, A_y X_z$ $A_z X_x, A_z X_y$	A spin x or y antiphase magnetization. Half of the magnetization lies along the positive axis and half of the magnetization lies along the negative axis. This distinguishes between 1–3 transitions and 2–4 transitions. This magnetization is unobservable but can be converted back into observable magnetization.
$A_z X_z$	Antiphase z magnetization, also known as a j ordered spin state, provides a polarization transfer pathway. This magnetization is unobservable but can be converted back into observable magnetization.

I.C.8. Density Matrix Treatment of DEPT

In this section, we will use some of the density matrix methods developed in the preceding sections to describe Distortionless Enhanced Polarization Transfer (DEPT) for an AX (A-proton and X-^{13}C) system.

We will use consistently the notation that in the product spin wave functions such as $\alpha\beta$, the first single particle function refers to the A (proton) nucleus, and the second to the X (^{13}C) nucleus. Thus, $\alpha\beta$ refers to the proton in the spin up state and the ^{13}C in the spin down state. We will write (see Eq. I.A.15),

$$\omega_A = 2\pi f_A = \gamma_A B_0, \qquad \omega_X = 2\pi f_X = \gamma_X B_0, \qquad \text{(I.C.105)}$$

and J for the coupling constant between A and X. J, f_A, and f_B will be expressed in Hz.

The energy level system appears as Fig. I.C.1. The Hamiltonian, expressed in units of h, is

$$H/h = \frac{1}{2}\begin{pmatrix} -f_A - f_X + J/2 & 0 & 0 & 0 \\ 0 & -f_A + f_X - J/2 & 0 & 0 \\ 0 & 0 & +f_A - f_X - J/2 & 0 \\ 0 & 0 & 0 & +f_A + f_B + J/2 \end{pmatrix}. \qquad \text{(I.C.106)}$$

If we write

$$f_A = 4f_X, \qquad \text{(I.C.107)}$$

which is an approximate empirical fact, and

$$p = h f_X/4kT, \qquad \text{(I.C.108)}$$

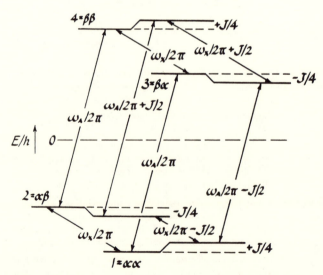

Figure I.C.1 Energy level diagram for an A–X system with J coupling. Separations are expressed in: frequency $\approx$ (angular frequency)$(2\pi)^{-1} \approx$ (energy) h^{-1}.

then

$$\rho(0) \approx p \begin{pmatrix} 5 & 0 & 0 & 0 \\ 0 & 3 & 0 & 0 \\ 0 & 0 & -3 & 0 \\ 0 & 0 & 0 & -5 \end{pmatrix}, \qquad (\text{I.C.109})$$

where we have neglected J in comparison with f_X.

Moreover, in the double rotating frame, in which the proton and ^{13}C frames rotate at angular frequencies ω_A and ω_X, we have, from Eq. I.C.57,

$$U_H = \exp[(i/\hbar)Ht] = \exp[(2\pi i)(H/h)t]. \qquad (\text{I.C.110})$$

This leads to

$$U_H = \begin{pmatrix} e^{-2\pi iJt/4} & 0 & 0 & 0 \\ 0 & e^{+2\pi iJt/4} & 0 & 0 \\ 0 & 0 & e^{+2\pi iJt/4} & 0 \\ 0 & 0 & 0 & e^{-2\pi iJt/4} \end{pmatrix}$$

$$= \begin{pmatrix} e^{-i\theta} & 0 & 0 & 0 \\ 0 & e^{+i\theta} & 0 & 0 \\ 0 & 0 & e^{+i\theta} & 0 \\ 0 & 0 & 0 & e^{-i\theta} \end{pmatrix}, \qquad (\text{I.C.111})$$

where we have written

$$\theta = 2\pi Jt/4. \qquad (\text{I.C.112})$$

The initial populations can be found by using

$$A_z = \tfrac{1}{2}\sigma_z \otimes \sigma_0 \qquad (\text{I.C.113})$$

(see Section I.C.7), and Eq. I.C.35:

$$A_0 = \text{Tr } \rho(0)A_z = 8p; \qquad X_0 = \text{Tr } \rho(0)X_z = 2p. \qquad (\text{I.C.114})$$

Let us now subject the AX system to a sequence of pulses that will lead to DEPT (Distortionless Enhancement of Polarization Transfer). The sequences at the resonance frequencies of both A and X appear as Fig. I.C.2. The scaled time θ is defined in Eq. I.C.112.

Use of Table I.C.1 leads to

$$\rho(0) = p \begin{pmatrix} 5 & 0 & 0 & 0 \\ 0 & 3 & 0 & 0 \\ 0 & 0 & -3 & 0 \\ 0 & 0 & 0 & -5 \end{pmatrix} \qquad \begin{aligned} \langle \bar{A}_z \rangle &= 8p = A_0 & \langle \bar{X}_z \rangle &= 2p = X_0 \\ \langle \bar{A}_+ \rangle &= 0 = \langle \bar{X}_+ \rangle \end{aligned}$$

$$(\text{I.C.115})$$

$$\rho(a) = p \begin{pmatrix} 1 & 0 & -4i & 0 \\ 0 & -1 & 0 & -4i \\ 4i & 0 & 1 & 0 \\ 0 & 4i & 0 & -1 \end{pmatrix} \qquad \begin{aligned} \langle \bar{A}_z \rangle &= 0 \quad \langle \bar{X}_z \rangle = X_0 \\ \langle \bar{A}_+ \rangle &= iA_0 \quad \langle \bar{X}_+ \rangle = 0 \end{aligned}$$

(I.C.115)

$$\rho(b) = p \begin{pmatrix} 1 & 0 & -4 & 0 \\ 0 & -1 & 0 & 4 \\ -4 & 0 & 1 & 0 \\ 0 & 4 & 0 & -1 \end{pmatrix} \qquad \begin{aligned} \langle \bar{A}_z \rangle &= 0 \quad \langle \bar{X}_z \rangle = X_0 \\ \langle \bar{A}_+ \rangle &= 0 \quad \langle \bar{X}_+ \rangle = 0. \end{aligned}$$

Up to this point, the density matrix approach yields nothing beyond what a semiclassical vector approach would. After the pulses at $\theta = \pi/4$, we have

$$\rho(c) = p \begin{pmatrix} 0 & -i & 0 & 4i \\ i & 0 & -4i & 0 \\ 0 & 4i & 0 & -i \\ -4i & 0 & i & 0 \end{pmatrix} \qquad \begin{aligned} \langle \bar{A}_z \rangle &= 0 = \langle \bar{X}_z \rangle \\ \langle \bar{A}^+ \rangle &= 0 \quad \langle \bar{X}_+ \rangle \times iX_0. \end{aligned}$$

(I.C.116)

Note that at this point the density matrix method predicts the counter-diagonal terms (on which the vector model is silent), as well as the observables. Zero quantum and two quantum coherences are described.

Proceeding, we obtain

$$\rho(d) = p \begin{pmatrix} 0 & -1 & 0 & 4i \\ -1 & 0 & -4i & 0 \\ 0 & 4i & 0 & 1 \\ -4i & 0 & 1 & 0 \end{pmatrix} \qquad \begin{aligned} \langle \bar{A}_z \rangle &= 0 = \langle \bar{X}_z \rangle \\ \langle \bar{A}_+ \rangle &= 0 = \langle \bar{X}_+ \rangle, \end{aligned}$$

(I.C.117)

which is a result not contained in the vector picture.

DEPT

$A(^1H)$: $(-90_x)_A - \theta = \pi/4 - (-180_x)_A - \theta = \pi/4 - (90_y)_A$

$X(^{13}C)$: $\qquad\qquad (-90_x)_x - \theta = \pi/4 - (-180_x)_x - \theta = \pi/4 - AQ$

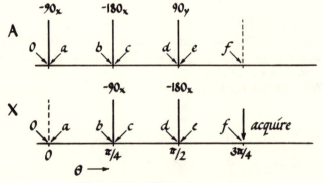

Figure 1.C.2 DEPT pulse sequence.

Taking the next step, we arrive at

$$\rho(e) = p \begin{pmatrix} 0 & 4i & 0 & -1 \\ -4i & 0 & -1 & 0 \\ 0 & -1 & 0 & -4i \\ -1 & 0 & 4i & 0 \end{pmatrix} \qquad \begin{aligned} \langle \bar{A}_z \rangle &= 0 = \langle \bar{X}_z \rangle \\ \langle \bar{A}_+ \rangle &= 0 = \langle \bar{X}_+ \rangle. \end{aligned} \qquad \text{(I.C.118)}$$

This leads, finally, to

$$\rho(f) = p \begin{pmatrix} 0 & 4 & 0 & -1 \\ 4 & 0 & -1 & 0 \\ 0 & -1 & 0 & 4 \\ -1 & 0 & 4 & 0 \end{pmatrix} \qquad \begin{aligned} \langle \bar{A}_z \rangle &= 0 = \langle \bar{X}_z \rangle \\ \langle \bar{A}_+ \rangle &= 0 \quad \langle \bar{X}_+ \rangle = 8p = A_0. \end{aligned}$$

$$\text{(I.C.119)}$$

indicating a transfer of polarization from the A system to the X system, with some residual ZQC and DQC (see Eq. I.C.104).

I.C.9. Graphical Portrayal of Basis Matrices

The somewhat abstract matrices are given a useful visualizable representation in the reference by Shriver [1992], at the end of this chapter.

I.D. NUCLEAR SPIN INTERACTIONS AND THE SOLOMON EQUATIONS

In Section I.B we treated the macroscopic effects of nuclear magnetization using the Bloch Equations. In this section, we will use the Solomon Equations to describe the effects of interactions among the nuclear magnetic moments themselves, beginning with the two-spin case, and generalizing to the case of many mutually interacting spins. These considerations are particularly important in understanding cross polarization, a fundamental technique in NMR in the solid state.

I.D.1. The Two-Spin Case

Consider a pair of isolated spin 1/2 nuclei not necessarily identical, interacting with one another. Spin I alone would have two energy levels separated by a $\hbar\omega_I$, and spin S would correspondingly have two energy levels separated by $\hbar\omega_S$. Since the two particles are interacting, however, we have a four level system, as shown in Fig. I.D.1. We write the population of the four levels as N_{++}, N_{+-}, N_{-+}, and N_{--} (the first subscript refers to I, the second, S) and write transition probabilities per unit time between the states as W_0, W_1, W'_1, and W_2 (as labeled in Fig. I.D.1) due to the interaction between the nuclear magnetic moments. The time rates of change of the populations of the four

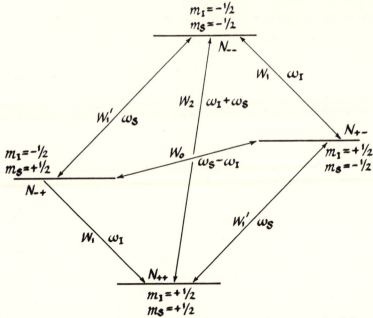

Figure I.D.1   Transition probabilities and populations in a four-energy-level diagram.

states satisfy

$$dN_{++}/dt = -(W_1 + W_1' + W_2)N_{++} + W_1'N_{+-} + W_1N_{-+} + W_2N_{--} + \text{const.},$$

(I.D.1)

$$dN_{+-}/dt = W_1' + N_{++} - (W_1' + W_1 + W_0)N_{+-} + W_0N_{-+} + W_1N_{--} + \text{const.},$$

(I.D.2)

$$dN_{-+}/dt = W_1N_{++} + W_0N_{+-} - (W_1' + W_1 + W_0)N_{-+} + W_1'N_{--} + \text{const.},$$

(I.D.3)

and

$$dN_{--}/dt = W_2N_{++} + W_1N_{+-} + W_1'N_{-+} - (W_1' + W_1 + W_2)N_{--} + \text{const.}$$

(I.D.4)

These equations follow simply from the meanings of the Ws; the constants appear to guarantee the Boltzmann equilibrium distribution of spins among the states.

The macroscopic magnetic moments I_z and S_z will be proportional to the excess populations in the spin up states of the respective nuclei. That is

$$(N_{++} + N_{+-}) - (N_{-+} + N_{--}) = kI_z$$

(I.D.5)

and

$$(N_{++} + N_{-+}) - (N_{+-} + N_{--}) = kS_z,$$

(I.D.6)

where k is a constant of proportionality.

Combining the last several equations, we obtain

$$dI_z/dt = -(W_0 + 2W_1 + W_2)I_z - (W_2 - W_0)S_z + \text{const.} \quad \text{(I.D.7)}$$

and

$$dS_z/dt = -(W_2 - W_0)I_z - (W_0 + 2W_1' + W_2)S_z + \text{const.} \quad \text{(I.D.8)}$$

We can determine the constants by observing that the populations should stop changing when $S_z = S_0$ and $I_z = I_0$, the thermal equilibrium values. Following this procedure, we obtain

$$dI_z/dt = -(W_0 + 2W_1 + W_2)(I_z - I_0) - (W_2 - W_0)(S_z - S_0) \quad \text{(I.D.9)}$$

and

$$dS_z/dt = -(W_2 - W_0)(I_z - I_0) - (W_0 + 2W_1' + W_2)(S_z - S_0) \quad \text{(I.D.10)}$$

Consider several special cases.

a. Suppose I and S are like spins; that is, suppose $\gamma_I = \gamma_S$. Under these circumstances, we can observe only $I_z + S_z$, not the spins separately. We therefore add the equations, noting that $W_1 = W_1'$ for this case:

$$dI_z/dt + dS_z/dt = d(I_z + S_z)/dt = -(2W_1 + 2W_2)[(I_z + S_z) + (I_0 + S_0)],$$
$$\text{(I.D.11)}$$

or

$$d(I_z + S_z)/dt = [(I_0 + S_0) - (I_z + S_z)]/T_1, \quad \text{(I.D.12)}$$

where we have written

$$1/T_1 = 2(W_1 + W_2), \quad \text{(I.D.13)}$$

where T_1 has the usual meaning of the time constant of the reapproach of the z-component of the magnetization to its thermal equilibrium value. The similarity of these equations to part of the Bloch Equations (see Section I.B) is clear.

b. Now suppose that I refers to a nuclear spin and S refers to an electronic spin. In this case, since electronic relaxation is very fast, we always have $S_z \approx S_0$. This leads to

$$dI_z/dt = -(W_0 + 2W_1 + W_2)(I_z - I_0) = (I_0 - I_z)/T_1, \quad \text{(I.D.14)}$$

where

$$1/T_1 = W_0 + 2W_1 + W_2. \quad \text{(I.D.15)}$$

c. Next, suppose I and S refer to unlike nuclei: $\gamma_j \neq \gamma_S$.
   If we define

$$\rho = W_0 + 2W_1 + W_2, \quad \text{(I.D.16)}$$

$$\rho_1' = W_0 + 2W_1' + W_2, \quad \text{(I.D.17)}$$

and

$$\sigma = W_2 - W_0. \quad \text{(I.D.18)}$$

We obtain

$$dI_z/dt = -\rho(I_z - I_0) - \sigma(S_z - S_0) \quad \text{(I.D.19)}$$

and

$$dS_z/dt = -\rho'(S_z - S_0) - \sigma(I_z - I_0). \quad \text{(I.D.20)}$$

Let us now suppose that the S spins are saturated ($S_z = 0$), and look for the steady state condition of the I spins:

$$dI_z/dt = 0 = -\rho(I_z - I_0) - \sigma(S_z - S_0) = -\rho(I_z - I_0) + \sigma S_0. \quad \text{(I.D.21)}$$

We have, then,

$$\rho(I_z - I_0) = \sigma S_0 \quad \text{(I.D.22)}$$

so

$$I_z = I_0 + (\sigma/\rho)S_0. \quad \text{(I.D.23)}$$

The enhancement of the I-spin polarization under these conditions is called the *Overhauser effect*.

Now back to general considerations. For notational simplicity, we will drop the subscript z, and will consider it understood that we are referring to the component of the spins in the direction of the external magnetic field. We have, then, for the two spin case,

$$-dI/dt = \rho(I - I_0) + \sigma(S - S_0) \quad \text{(I.D.24)}$$

and

$$-dS/dt = \rho'(S - S_0) + \sigma(I - I_0), \quad \text{(I.D.25)}$$

where ρ_1, ρ'_1, and σ are defined as in Eqs. I.D.16–I.D.18.

I.D.2. The Transition Probabilities and T_1

In order to calculate the transition probabilities appearing in the Solomon equations, we will write down the Hamiltonian for the system:

$$H = H_M - \hbar\gamma_I B_0 I_z - \hbar\gamma_S B_0 S_z + H'. \quad \text{(I.D.26)}$$

In this equation, the second and third terms represent the Zeeman energies of the I and S spins in the strong external magnetic field $\mathbf{B}_0$, presumed to be in the z-direction, and H_M represents all contributions to the energy of the system that do not depend on the spin coordinates (the "lattice"). H' is a term that depends on both I and S spins, as well as the lattice: it is the interaction term between the spins I and S. Typically, for the cases we will consider, H' is small in comparison with the other terms in Eq. I.D.45, and receives a time-dependence through the coordinates of the lattice. One uses standard time-dependent perturbation theory or density matrix theory (see Section I.C) to calculate transition probabilities due to H' between the Zeeman energy states of I and S.

The interaction term may represent many types of interaction, such as indirect, spin–spin coupling, a quadrupolar interaction, or a spin–rotation interaction. We will choose as an important example the direct dipole–dipole interaction.

The form of the dipole–dipole interaction is (see Section III.A):

$$H' = -(\hbar^2\gamma_I\gamma_S/r^3)[3(\mathbf{I}\cdot\hat{\mathbf{r}})(\mathbf{S}\cdot\hat{\mathbf{r}}) - \mathbf{I}\cdot\mathbf{S}], \quad \text{(I.D.27)}$$

where $\mathbf{r}$ is the internuclear vector, and $\hat{\mathbf{r}}$ is the unit vector in the direction of **r**,

$$\hat{\mathbf{r}} = \mathbf{r}/r. \tag{I.D.28}$$

We will take as the unperturbed spin energy states direct products of the eigenstates of I_z and S_z, which, as is clear from Eq. I.D.45, are also eigenstates of the Zeeman Hamiltonian:

$$I_z|+\rangle = +(1/2)|+\rangle, \tag{I.D.29}$$

$$I_z|-\rangle = -(1/2)|-\rangle, \tag{I.D.30}$$

$$S_z|+) = +(1/2)|+), \tag{I.D.31}$$

and

$$S_z|-) = -(1/2)|-), \tag{I.D.32}$$

so that the four unperturbed eigenfunctions are

$$|1\rangle = |+\rangle|+), \tag{I.D.33}$$

$$|2\rangle = |+\rangle|-), \tag{I.D.34}$$

$$|3\rangle = |-\rangle|+), \tag{I.D.35}$$

and

$$|4\rangle = |-\rangle|-). \tag{I.D.36}$$

H' receives its time-dependence from $\mathbf{r}(t)$, which can vary in magnitude as **I** and **S** move closer or further apart, or can vary in direction even if the nuclei are fixed at a certain distance in a molecule which is rotating.

A fundamental result of time-dependent perturbation theory (see Section I.C.3) is that the transition probabilities per unit time from state i to state j induced by the perturbation $H'(t)$ is proportional to the absolute value squared of the matrix element of the perturbation between initial and final states:

$$W_{ij} = (1/\hbar^2 t)\left|\int_0^t \langle m_i|H'(t')|m_j\rangle e^{-i\omega t'_j}\,dt'\right|^2. \tag{I.D.37}$$

Different terms in the expression for the dipole–dipole interaction given in Section III.A.1 will have non-vanishing matrix elements between different initial and final states: for example, the term E in Eq. III.A.11 will have a non-vanishing matrix element for states $m_i = |4\rangle$ and $m_j = |1\rangle$. We will work out W_{ij} for this special case, which along with the term F in Eq. III.A.11 corresponds to W_2 (see Fig. I.D.1).

Using Appendix A1, we can show

$$I_+|-\tfrac{1}{2}\rangle = (I_x + iI_y)|-\tfrac{1}{2}\rangle = 2(\sigma_x + \sigma_y)|-\tfrac{1}{2}\rangle = |+\tfrac{1}{2}\rangle, \tag{I.D.38}$$

so

$$\langle 1|I_+S_+|4\rangle = \langle 1|1\rangle = 1; \tag{I.D.39}$$

therefore

$$E = -(\tfrac{3}{4})\sin^2\theta\, e^{-i2\phi}[I_+S_+] \tag{I.D.40}$$

leads to

$$\langle 1|E|4\rangle = -(\tfrac{3}{4})\sin^2\theta(t)\, e^{-i2\phi(t)} = E_2(t)/\hbar\omega_D, \tag{I.D.41}$$

which defines $E_2(t)$; the subscript 2 reminds us that this refers to W_2, and ω_D is defined in Eq. III.A.10. We have

$$W_{14} = (1/\hbar^2 t)\left|\int_0^t E_2(t') \exp(-i[\omega_I + \omega_S]t')\,dt'\right|^2, \qquad \text{(I.D.42)}$$

since

$$\omega_{14} = (E_1 - E_4)/\hbar = \omega_I + \omega_S \qquad \text{(I.D.43)}$$

where

$$\omega_I = \gamma_I B_0 \qquad \text{and} \qquad \omega_S = \gamma_S B_0. \qquad \text{(I.D.44)}$$

How do we evaluate the integral appearing in Eq. I.D.42? $E_2(t')$ receives its time-dependence through the spherical polar coordinates of $r(t)$, which we will assume are random functions of time. Such random functions are typically treated in terms of correlation functions, which express the rate at which the original functions lose memory of their earlier values. To use this approach, we need to rewrite Eq. I.D.42 in the form.

$$\langle W_{14}\rangle = \langle W_2\rangle = (1/\hbar)\int_{-\infty}^{+\infty} d\tau \exp[-i(\omega_I + \omega_S)\tau]K_2(\tau) \qquad \text{(I.D.45)}$$

where

$$K_2(\tau) = \langle E_2(t')E_2^*(t' - \tau)\rangle. \qquad \text{(I.D.46)}$$

To make a connection between Eqs. I.D.42 and I.D.45, rewrite Eq. I.D.42 as follows:

$$W_{14} = (1/\hbar t)\int_0^t E_2(t') \exp(-i[\omega_I + \omega_S]t')\,dt'\int_0^t E_2^*(t'') \exp(i[\omega_I + \omega_S]t'')\,dt''$$

$$\text{(I.D.47)}$$

Combining the integrals, we obtain

$$W_{14} = (1/\hbar^2 t)\int_0^t\int_0^t E_2(t')E^*(t') \exp[-i(\omega_I + \omega_S)(t' - t'')]\,dt'\,dt''. \quad \text{(I.D.48)}$$

Now define τ by

$$\tau = t' - t''. \qquad \text{(I.D.49)}$$

This change of variables allows us to rewrite the integral as

$$W_{14} = (1/\hbar^2 t)\left\{\int_0^t d\tau \exp[-i(\omega_I + \omega_S)\tau]\int_\tau^t E_2(t')E_2^*(t' - \tau)\,dt'\right.$$
$$\left. + \int_\tau^0 d\tau \exp[-i(\omega_I + \omega_S)\tau]\int_0^{t+\tau} E_2(t')E_2^*(t' - \tau)\,dt'\right\}. \quad \text{(I.D.50)}$$

Using the definition of correlation functions of E_2 as

$$K_2(\tau) = \langle E_2(t')E_2^*(t' - \tau)\rangle, \qquad \text{(I.D.51)}$$

we can evaluate the integrals over t' by taking the average value of the integrand times the range between the integration limits, so that

$$\langle W_2 \rangle = \langle W_{14} \rangle = (1/\hbar^2 t)\left\{ \int_0^t d\tau \exp[-i(\omega_I + \omega_S)\tau](t - \tau)K_2(\tau) \right.$$

$$\left. + \int_{-t}^0 d\tau \exp[-i(\omega_I + \omega_S)\tau](t + \tau)K_2(\tau) \right\} \qquad (I.D.52)$$

$$= (1/\hbar^2 t)\left\{ t \int_{-t}^t d\tau \exp[-i(\omega_I + \omega_S)\tau]K_2(\tau) \right.$$

$$\left. - 2\int_0^t d\tau \cos[(\omega_I + \omega_S)\tau]\tau K_2(\tau) \right\}. \qquad (I.D.53)$$

Using Eq. I.D.54, we can see that the magnitude of the ratio of the two integrals in Eq. I.D.53 is of the order of t/τ_c. This will be quite large in a typical NMR experiment: it will be related to the ratio of the relaxation time to the correlation time, which, for the case of molecular rotation, is of the order of the time required for the molecule to rotate through a radian. We neglect, then, the second integral, and, for the same reason, extend the limits on the first integral to $\pm\infty$, since the integrand is essentially zero for times greater than t or less than zero.

One usually assumes an exponential decay as a function of τ:

$$K_2(\tau) = K_2(0) \exp(-|\tau|/\tau_c), \qquad (I.D.54)$$

where Eq. I.D.54 defines the correlation time τ_c.

The Fourier transform of the correlation function extracts the frequency dependence of the fluctuating variable:

$$J_2(\omega) = \int_{-\infty}^\infty K_2(\tau) \exp(-i\omega t) \, d\tau = K_2(0)2\tau_c/(1 + \omega^2\tau_c^2) \qquad (I.D.55)$$

from the standard Fourier transform of an exponentially decaying function.

Fig. I.D.2 shows a plot of $J_2(\omega)$. T_1 is inversely proportional to $J(\omega)$, which

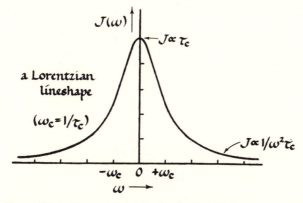

Figure I.D.2 Spectral intensity distribution; Debye spectrum.

represents the spectrum of frequencies of fluctuating magnetic fields at the site of a specific nuclear dipole arising from nearby dipoles. Physically, the larger this spectrum is at the resonance frequency, the more effective the relaxation mechanism, and the shorter the relaxation time.

By definition, we have

$$K2(0) = \langle E2(t)E2^*(t) \rangle. \tag{I.D.56}$$

We will use a simple, historically important, special case to evaluate Eq. I.D.56: suppose $\mathbf{I}$ and $\mathbf{S}$ are fixed at a constant distance r, and the time-dependence occurs in the spherical polar coordinates θ and φ as the molecule tumbles randomly.

For this case,

$$K_2(0) = (4\pi)^{-1} \int_0^{2\pi} \int_0^{\pi} E_2 E_2^* \sin\theta \, d\theta \, d\phi$$

$$= (4\pi)^{-1} \hbar^2 \omega_D^2 \int_0^{2\pi} \int_0^{\pi} \sin^5\theta \, d\theta \, d\phi = (8/15)\hbar^2 \omega_D^2. \tag{I.D.57}$$

We have, then,

$$\langle W_2 \rangle = (1/\hbar^2)J_2(\omega_I + \omega_S). \tag{I.D.58}$$

An explicit expression for $\langle W_2 \rangle$ can be found for the case of rigid random rotation by evaluating $J_2(\omega_I + \omega_S)$ using Eqs. I.D.55 and I.D.57. This leads to

$$\langle W_2 \rangle = (3\hbar^2\gamma_I^2\gamma_S^2/20r^6)\{\tau_c/[1 + (\omega_I + \omega_S)^2\tau_c^2]\}. \tag{I.D.59}$$

Similar analyses lead to expressions for $\langle W_1 \rangle$, $\langle W_1' \rangle$, and $\langle W_0 \rangle$.

Putting them all together, one obtains, for the case of two like spins ($\gamma_I = \gamma_S$), Eq. I.D.13,

$$T_1^{-1} = (3\hbar^2\gamma^4/10r^6)\{\tau_c/(1 + \omega^2\tau_c^2) + 4\tau_c/(1 + 4\omega^2\tau_c^2\}. \tag{I.D.60}$$

We can write Eq. I.D.60 as

$$(T_1)^{-1} = C\{\tau_c/(1 + \omega^2\tau_c^2) + 4\tau_c/(1 + 4\omega^2\tau_c^2)\}. \tag{I.D.61}$$

Let us fix ω and vary τ_c and look at the variation of T_1. Typically, we can vary τ_c by varying the temperature T, since, under most circumstances, we have

$$\tau_c = \tau_0 \exp(-E_a/kT), \tag{I.D.62}$$

where E_a is an activation energy. We will begin by considering two limiting values of Eq. I.D.61:

If

$$\omega^2\tau_c^2 \ll 1, \tag{I.D.63}$$

then

$$(T_1)^{-1} = 5C\tau_c. \tag{I.D.64}$$

If

$$[\omega^2\tau_C^2 \gg 1], \tag{I.D.65}$$

then

$$(T_1^{-1}) = 2C/(\omega^2\tau_c). \tag{I.D.66}$$

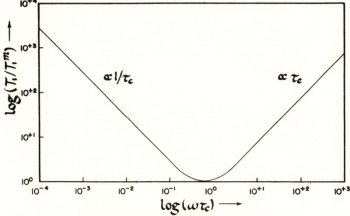

Figure I.D.3 Dependence of T_1 on τ_c.

We can differentiate Eq. I.D.61 with respect to τ_c and find that the minimum occurs at

$$T_1^m = \omega/1.4252C \quad \text{for } \tau_c^m = 0.61579/\omega. \qquad (\text{I.D.67})$$

Guided by these considerations, we can sketch $\log T_1$ as a function of $\log \tau_c$, or, using Eq. I.D.62, as a function of $1/T$ (see Fig. I.D.3). The relaxation time passes through a minimum value as the temperature (and hence the correlation time) is varied.

Conversely, we can fix τ_c and observe the variation of the relaxation as a function of frequency ω. Reference to Eq. I.D.61 is enough to convince one that in the low frequency limit ($\omega \to 0$), T_1 approaches a constant value $(5C\tau_c)^{-1}$, whereas at the high frequency limit ($\omega^2\tau_c^2 \gg 1$), T_1 is proportional to ω^2 (see Fig. I.D.4).

This analysis is intended to be only an elementary introduction to the question of calculating relaxation times in NMR experiments. The details of the relaxation mechanism affect the equations for the relaxation times, but the outline of the method is, to a certain extent, generalizable.

Historically, the first calculation of relaxation times dealt with water, and the relaxation mechanism was the intramolecular dipole–dipole interaction between the two protons, considered as being fixed in a spherical molecule tumbling randomly in a viscous fluid. For this case, Debye's expression for the correlation time as a function of temperature could be used:

$$\tau_c = 4\pi\eta a^3/3kT, \qquad (\text{I.D.68})$$

where a is the radius of the water molecule, and η is the viscosity.

I.D.3. The Transverse Relaxation Time T_2

As discussed in Chapter I, the relaxation time for the component of nuclear magnetization that is transverse to the external magnetic field, T_2, is not necessarily the same as the longitudinal relaxation time, T_1, because the

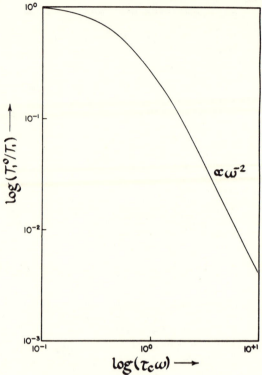

Figure I.D.4 Dependence of T_1 on ω.

latter involves an exchange of energy between the spin system with the lattice, while the former does not. This is illustrated by the special case we have been using as an example, protons in water.

It can be shown for this case, using either time-dependent perturbation or density matrix theory, that the dipole–dipole interaction leads to

$$(T_2)^{-1} = (3/20)(h^2\gamma^4/r^6)\{3\tau_c + 5\tau_c/(1 + \omega^2\tau_c^2) + 2\tau_c/(1 + 4\omega^2\tau_c^2)\}. \qquad \text{(I.D.69)}$$

For the limiting case

$$\omega^2\tau_c^2 \ll 1, \qquad \text{(I.D.70)}$$

we have

$$(T_2)^{-1} = (3/20)(\hbar^2\gamma^4/r^6)(10\tau_c). \qquad \text{(I.D.71)}$$

Reference to Eq. I.D.64 shows that, in this limit, $T_1 = T_2$.

On the other hand, when $\omega^2\tau_c^2 \gg 1$,

$$(T_2)^{-1} = (3/20)(\hbar^2\gamma^4/r^6)(3\tau_c), \qquad \text{(I.D.72)}$$

which, as can be seen from Eq. I.D.66, differs from T_1.

For very long τ_c, corresponding (through Eq. I.D.50) to a low temperature, the analysis breaks down because the macroscopic system is no longer describable as a Deybe liquid. At $\tau_c \approx T_2$, T_2 approaches a constant value asymptotically (see Fig. I.D.5).

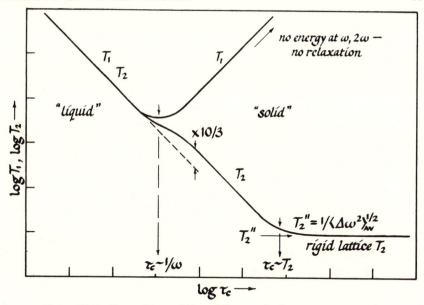

Figure I.D.5 Dependence of T_1 and T_2 on τ_c (or inverse temperature).

I.D.4. Other Relaxation Mechanisms

We have treated here only a single relaxation mechanism, the dipole–dipole interaction. Other relaxation mechanisms are:

a. Quadrupole interaction for $I > \frac{1}{2}$.

$$H_Q = \mathbf{I} \cdot \mathbf{Q} \cdot \mathbf{I}, \qquad (\text{I.D.73})$$

where $\mathbf{Q}$, the quadrupole coupling tensor, fluctuates with molecular motion and H_Q induces transitions between energy levels.

b. Chemical Shift Anisotropy.

The local magnetic field at the site of a nucleus is

$$\mathbf{B}_{\text{loc}} = \mathbf{B}_0(1 - \sigma), \qquad (\text{I.D.74})$$

where $\mathbf{B}_0$ is the external magnetic field and σ, defined by this equation, is the chemical shift parameter (see Sections I.A.5 and IV.A).

The chemical shift part of the Hamiltonian is given by

$$H_{CS} = -\gamma \hbar \mathbf{B}_0 \cdot \sigma \cdot \mathbf{I}, \qquad (\text{I.D.75})$$

If the chemical shift tensor is anisotropic, then, as molecular motion occurs, the components of the chemical shift tensor σ fluctuate, giving rise to nuclear spin energy level transitions.

c. Scalar Relaxation (J coupling).

Two spins $\mathbf{I}$ and $\mathbf{S}$ can couple through an electron system. The Hamiltonian has the form

$$H_S = \hbar \mathbf{I} \cdot \mathbf{A} \cdot \mathbf{S}. \qquad (\text{I.D.76})$$

In liquids, this takes the form of a scalar spin–spin interaction (see Section I.A.5).

d. Spin–Rotation Interaction.

A nucleus can couple with the magnetic moment of a molecule related to the total angular momentum of the molecule. The interaction is called the spin–rotation interaction, and has the form

$$H_{SR} = \hbar \mathbf{I} \cdot \mathbf{C} \cdot \mathbf{J}, \qquad (I.D.77)$$

where $\mathbf{J}$ is the angular momentum operator for molecular rotation and $\mathbf{C}$ is the spin–rotation tensor.

In each of these cases, the interaction Hamiltonian depends on the nuclear spin coordinates of the nucleus and receives a time-dependence through the presence of coordinates of the lattice. Relaxation times due to these interactions can be calculated using methods outlined in our treatment of the dipole–dipole interaction, using either time-dependent perturbation theory or density matrix theory.

I.D.5. The Density Matrix and Relaxation

We will now present an introduction to an alternative approach to the problem of nuclear magnetic relaxation, based on the density matrix (see Section I.C.4).

One considers a system described by a Hamiltonian that can be divided according to

$$H(t) = H_0 + H_1(t), \qquad (I.D.78)$$

where H_0 is time-*independent*, and $H_1(t)$ is, in some sense, much smaller than H_0. Examples from magnetic resonance would be when H_0 is the Zeeman Hamiltonian of a spin system in a strong external magnetic field and $H_1(t)$ is the coupling of the spins to a small, oscillating field, or a spin interaction term such as the dipole coupling between two spins that receives its time-dependence through molecular motion.

Eq. I.C.56 for the density matrix becomes, in this case,

$$\frac{d\rho}{dt} = \frac{i}{\hbar}[\rho, H_0 + H_1(t)]. \qquad (I.D.79)$$

If H_1 is zero, we can see by substitution that a solution for $\rho(t)$ is

$$\rho(t) = \exp(-iH_0 t/\hbar)\rho(0)\exp(iH_0 t/\hbar). \qquad (I.D.80)$$

With this as a guide, let us define a reduced density matrix $\sigma(t)$ by

$$\sigma(t) = \exp(-iH_0 t/\hbar)\sigma(t)\exp(iH_0 t/\hbar). \qquad (I.D.81)$$

Let us also define an operator $h(t)$ by

$$h = \exp(iH_0 t/\hbar)H_1(t)\exp(-iH_0 t/\hbar). \qquad (I.D.82)$$

Substitution and some algebra lead to

$$\frac{d\sigma}{dt} = \frac{i}{\hbar}[\sigma, h]. \tag{I.D.83}$$

So far, the calculation is exact.

We can write the time-dependent wavefunction $\Psi(t)$ as a linear combination of time-independent eigenfunctions of H_0 as follows:

$$\Psi(t) = \sum_n c_n(t)\phi_n \tag{I.D.84}$$

where

$$H_0\phi_n = E_n\phi_n, \tag{I.D.85}$$

and where the time-dependence of Ψ appears in the time-dependence of the expansion coefficients c_n (see Section I.C.3).

The equation of motion for $\sigma(t)$ can be integrated as follows:

$$\sigma(t) = \sigma(0) + \frac{i}{\hbar}\int_0^t [\sigma(t'), h(t')] \, dt'. \tag{I.D.86}$$

This integral can be approximated by an iterative procedure in which $\sigma(t')$ in the integrand is initially replaced by $\sigma(0)$,

$$\sigma_1(t) = \sigma(0) + \frac{i}{\hbar}\int_0^t [\sigma(0), h_1(t')] \, dt', \tag{I.D.87}$$

and then a better approximation obtained by inserting this expression into Eq. I.D.88:

$$\sigma_2(t) = \sigma(0) + \frac{i}{\hbar}\int_0^t [\sigma(0), h_1(t')] \, dt \tag{I.D.88}$$

$$+ \left(\frac{i}{\hbar}\right)^2 \int_0^t \int_0^t [[\sigma(0), h_1(t'')], h_1(t')] \, dt' \, dt''. \tag{I.D.89}$$

When higher order iterative terms are neglected (justified by the smallness of the relaxation Hamiltonian relative to the Zeeman energy), one can show, after considerable work, that

$$\frac{d\rho_{pq}}{dt} = \frac{i}{\hbar}[\rho, H_0]_{pq} + \sum_{r,s}' R_{pqrs}(\rho_{rs} - \rho_{rs0}), \tag{I.D.90}$$

where the sum in the second term includes only those terms for which

$$E_p - E_q = E_r - E_s, \tag{I.D.91}$$

where the R_{pqrs} involve matrix elements of the perturbing Hamiltonian and where ρ_{rs0} is the thermal equilibrium matrix element of the density matrix between states r and s.

Eq. I.D.90 is called the Redfield equation. Using this equation and

Eq. I.C.35, one can arrive at the time-dependence of the longitudinal and transverse components of the nuclear magnetization, and, hence, determine T_1 and T_2.

BIBLIOGRAPHY

Books

Abragam, A., *The Principles of Nuclear Magnetism*, Clarendon Press, Oxford, 1961. A scientific classic.

Andrew, E. R., *Nuclear Magnetic Resonance*, Cambridge University Press, 1955. The first NMR book?

Bloembergen, N., *Nuclear Magnetic Relaxation*, W. A. Benjamin, Inc., New York, NY, 1961. A reprint of his thesis; another classic.

Chandrakumar, N. and Subramanian, S., *Modern Techniques in High-Resolution FT-NMR*, Springer-Verlag, New York, NY, 1987.

Ernst, R. R., Bodenhausen, G., and Wokaun, A., *Principles of Nuclear Magnetic Resonance in One and Two Dimensions*, Clarendon Press, Oxford, 1986. Successor to Abragram?

Farrar, T. C. and Harriman, J. E., *Density Matrix Theory and its Applications in Spectroscopy*, Farragut Press, Madison, WI, 1989. See also: T. C. Farrar, *Concepts Magn. Reson.*, **2**:1 (1990); *ibid*, **2**:55 (1990). The best place to start density matrices.

Farrar, Thomas C. and Becker, Edwin, D., *Pulse and Fourier Transform NMR*, Academic Press, New York, 1971. Basic.

Farrar, Thomas C., *Pulsed Nuclear Magnetic Resonance Spectroscopy*, Farragut Press, Madison, WI, 1989. Update of the above; easy to take.

Feenberg, E. and Pake, G. E., *Notes on the Quantum Theory of Angular Momentum*, Addison-Wesley, Cambridge, MA, 1953. A minor classic.

Freeman, R., *A Handbook of Nuclear Magnetic Resonance*, John Wiley & Sons, New York, NY. 1987. Answers to (mostly) qualitative questions.

Gerstein, B C. and Dybowski, C. R., *Transient Techniques in NMR of Solids*, Academic Press, Orlando, FL, 1985. Especially multiple pulse decoupling.

Goldman, Maurice, *Spin Temperature and Nuclear Magnetic Resonance in Solids*, Clarendon Press, Oxford, 1970. A sequel to Abragam.

Goldstein, H., *Classical Mechanics*, Addison-Wesley, Cambridge, MA, 1950.

Harris, Robin K., *Nuclear Magnetic Resonance Spectroscopy*, Pitman Books, London, 1983. Chemist's book; physically sound.

Hohmans, S. W., *A Dictionary of NMR Concepts*, Clarendon Press, Oxford, 1989. Answers for the modern NMR spectroscopist's questions. Repetitious instead of ruthlessly cross referenced; as a result, easy to use.

Kaplan, J. I. and Fraenkel, G., *NMR of Chemically Exchanging Systems*, Academic Press, New York, NY, 1980.

Merhring, Michael, *High Resolution NMR in Solids*, 2nd Ed., Springer-Verlag, Berlin, 1983. Title is accurate. Out of print; a shame.

Sanders, J. K. M. and Hunter, B. K., *Modern NMR Spectroscopy: a guide for chemists*, Clarendon Press, Oxford, 1987. Mostly for bio/organic chemist users.

Slichter, C. P., *Principles of Magnetic Resonance*, 3rd Ed., Springer-Verlag, Berlin, 1989. (See also the first two editions.) By a master teacher.

Articles

I.A.1

Pake, G. E. (1950) "Fundamentals of Nuclear Magnetic Resonance Absorption. I," *Am. J. Phys.* **18**, 438–452.

Pake, G. E. (1950) "Fundamentals of Nuclear Magnetic Resonance Absorption. II," *Am. J. Phys.* **18**, 473–486.

I.B

Bloch, F. (1946) "Nuclear Induction," *Phys. Rev.* **70**, 460–474.

I.B.4

Hahn, E. L. (1950) "Nuclear Induction Due to Free Larmor Precession," *Phys. Rev.* **77**, 297–298.

Hahn, E. L. (1953) "Free Nuclear Induction," *Phys. Today* **6**: 11, 4–9.

I.B.6

Bloemberger, N., Purcell, E. M., and Pound, R. V. (1948) "Relaxation Effects in Nuclear Magnetic Resonance Absorption," *Phys. Rev.* **73**, 679–712.

I.B.7

Das, T. P. and Saha, A. K. (1954) "Mathematical Analysis of the Hahn Spin-Echo Experiment," *Phys. Rev.* **93**, 749–756.

Meiboom, S. and Gill, D. (1958) "Modified Spin-Echo Method for Measuring Nuclear Relaxation Times," *Rev. Sci. Instrum.* **29**, 688–691.

Stejskal, E. O. and Tanner, J. E. (1965) "Spin Diffusion Measurements: Spin Echoes in the Presence of a Time-Dependent Field Gradient," *J. Chem. Phys.* **42**, 288–292.

Stejskal, E. O. (1965) "Use of Spin Echoes in a Pulsed Magnetic-Field Gradient to Study Anisotropic, Restricted Diffusion and Flow," *J. Chem. Phys.* **43**, 3597–3603.

Woessner, D. E. (1961) "Effects of Diffusion in Nuclear Magnetic Resonance Spin-Echo Experiments," *J. Chem. Phys.* **34**, 2057–2061.

I.C.7

Farrar, T. C. (1990) "Density Matrices in NMR Spectroscopy: Part I," *Concepts Magn. Reson.* **2**, 1–12.

Farrar, T. C. (1990) "Density Matrices in NMR Spectroscopy: Part II," *Concepts Magn. Reson.* **2**, 55–61.

I.C.9

Shriver, J. (1992) "Product Operators and Coherence Transfer in Multiple-Pulse NMR Experiments," *Concepts Magn. Reson.* **4**, 1–33.

I.D

Bloemberger, N., Purcel, E. M., and Pound, R. V. (1948) "Relaxation Effects in Nuclear Magnetic Resonance Absorption," *Phys. Rev.* **73**, 679–712.

Kubo, R. and Tomita, K. (1954) "A General Theory of Magnetic Resonance Absorption," *J. Phys. Soc. Japan* **9**, 888–919.

Solomon, I. (1955) "Relaxation Processes in a System of Two Spins," *Phys. Rev.* **99**, 559–565.

I.D.1

Solomon, I. and Bloembergen, N. (1956) "Nuclear Magnetic Interactions in the HF Molecule," *J. Chem. Phys.* **25**, 261–66.

I.D.5

Hubbard, P. S. (1958) "Nuclear Magnetic Relaxation of Three and Four Spin Molecules in a Liquid," *Phys. Rev.* **109**, 1153–1158.

Hubbard, P. S. (1961) "Quantum-Mechanical and Semiclassical Forms of the Density Operator Theory of Relaxation," *Rev. Mod. Phys.* **33**, 249–264.

Redfield, A. G. (1957) "On the Theory of Relaxation Processes," *IBM Journal of Res. and Dev.* **1**, 19–31.

II

High Resolution Methods in the Solid State

The first section of this book deals with general principles of NMR; in this section, we will particularize to NMR of solid samples, with particular attention to techniques that permit high resolution studies to be done in the solid state. The first several parts of this section deal in a descriptive, qualitative fashion with important concepts. The solid state itself is described, with emphasis on how line broadening mechanisms such as direct dipole–dipole coupling and chemical shift anisotropy (CSA) due to the restricted motion of atoms and molecules in the solid state (in comparison with the liquid state) give rise to large linewidths. Special emphasis is given to organic solids in which the nuclear magnetic moments are protons and ^{13}C nuclei. Typically the protons will form a tightly coupled abundant spin system, while ^{13}C nuclei form a rare spin system since the natural abundance of the isotope is only about 1%.

There are also descriptive treatments of techniques to reduce the linewidth and make possible observation of hyperfine structure: both heteronuclear and homonuclear decoupling and magic angle spinning (MAS) (MAS is discussed in mathematical detail in Section V). There is also a theoretical treatment of cross polarization, which is a technique designed to enhance ^{13}C sensitivity through the dipole–dipole interaction between the ^{13}C spins and the proton spins.

II.A. DESCRIPTION OF THE SOLID STATE

Solid samples in NMR experiments exhibit the effect of the dipole–dipole interaction among the nuclei in a way not observed in liquid samples. In liquid samples, the molecules typically execute rapid and random rotation so that the interaction has no time-averaged effect (a detailed mathematical treatment of the dipole–dipole interaction appears in Section III). In solid samples, the atoms and molecules execute only restricted motion, so that the effect of the dipole–dipole interaction does not average to zero. Since the resonance frequency of a particular nucleus depends on the magnetic field at its site, and since the local field due to neighboring spins varies appreciably from place to place throughout the sample owing to a variation in the orientation of the neighboring spins, there will be a significant spread in the resonance frequencies, resulting in a line broader by several orders of magnitude than a typical line from a liquid sample.

The shape of the line is more or less Gaussian, which is described by the equation

$$G(\omega) = A \exp(-\omega^2/2\sigma^2), \qquad (II.A.1)$$

where ω is the angular frequency and σ is related to the width of the line. The line shape for a liquid sample, in addition to being much narrower, more closely approximates a Lorentzian, described by the equation

$$G(\omega) = \frac{A/T_2}{\omega^2 + (1/T_2)^2}, \qquad (II.A.2)$$

where T_2 is the spin–spin relaxation time.

NMR studies of solid samples which do not make use of high resolution techniques as discussed here are called "broadline" spectroscopy and the analysis of the experiment relates the moments of the line shape function to the physical parameters of the sample system. The nth moment is defined by the equation

$$M_n = \int_0^\infty (\omega - \omega_0)^n G(\omega)\, d\omega \bigg/ \int_0^\infty G(\omega)\, d\omega, \qquad (II.A.3)$$

where ω_0 is the centre of the line.

Another source of line broadening in the NMR of solid samples is chemical shift anisotropy (CSA). As discussed in Section I.A.5, a chemical shift in an NMR absorption line is the displacement of the resonance frequency due to a local field produced by a variation in the electron orbits in the strong external magnetic field necessary for the NMR experiment. In general the chemical shift is a tensor rather than a scalar, since the shift may depend on the orientation of the molecule with respect to the external field. In a polycrystalline sample, molecular orientation will vary from crystallite to crystallite, with the effect that there will be a continuous spread in the shifted lines, thus providing an additional source of broadening in the solid state. CSA is described mathematically in Section IV.B.2.

In an organic sample, the proton spin system is abundant and the ^{13}C spin system is rare, in the sense described previously. A given proton, that is, will interact through the dipole–dipole interaction with other protons, so the protons form a tightly coupled system. A ^{13}C nucleus, on the other hand, is typically far enough away from other ^{13}C so that this interaction is negligible. The ^{13}C nuclei thus are isolated with respect to other such nuclei, while interacting with several neighboring protons. This difference between the abundant and rare spin systems has important results, to be discussed later. In this presentation we will use the phrase "sparse spins" to describe a spin system in which the spins interact with each other weakly or not at all. The phrase "dilute spins" is also used to describe this situation although that suggests the deliberate act of dilution. Commonly, the phrase "rare spins" is used to describe such a case. However, that seems to require low isotopic abundance, which is not a necessary condition for the spins to be dilute or sparse. For example, while the isotopically rare spin ^{13}C is also sparse or dilute, so long as it is not enriched, the isotopically abundant spin ^{31}P is also usually sparse or dilute since most of its compounds separate the individual phosphorus atoms with other atoms, often oxygen atoms, which has the effect of seriously weakening direct P–P interactions.

II.B. DECOUPLING

One experimental technique to simplify the analysis of NMR spectra is *decoupling*, in which the interaction between nuclear spins, either the scalar interaction described in Section I.A.5 or the direct dipole–dipole interaction as described in Section III.A, is suppressed. Most of the earlier application of this technique is used in the high resolution study of liquids. Reference to Section I.A.5 and Fig. I.A.4, which describe the effect on the NMR spectrum of a scalar J coupling between two nuclei with different chemical shifts, will serve to illustrate the method in simple form. The splitting of the line of nucleus X due to its J coupling with nucleus A arises from the shift in the energy levels of nucleus X between the cases in which the coupling nucleus A has its spin up or down. If one can induce rapid (on a time scale related to the magnitude of the coupling constant) transitions between the up and down states of nucleus A, nucleus X sees a sort of average between the two states of nucleus A and the splitting of the X line collapses. These decoupling transitions are brought about by irradiation with radio frequency at the resonance frequency of the nuclear spin to the decoupled X. Early work using this technique was generally aimed both at simplifying complex high resolution spectra, and for identifying which nuclei were coupled by the J interaction. To illustrate this latter reason, one can decouple a particular nuclear spin and observe what part of the rest of the spectrum simplifies; only those multiplets representing nuclei coupled to that nuclear spin undergoing decoupling will simplify during the process.

This technique can also be used to narrow the broad lines arising from solid samples. Consider the broad line representing the sparse ^{13}C nuclear spins in an organic solid. Much of the linewidth is due to the dipole–dipole interaction of the ^{13}C nuclear spin with the sea of the abundant proton spins. While clearly related to decoupling in the liquid state, decoupling in the solid state differs experimentally in that the rf power in the decoupling radiation must be several orders of magnitude higher in order to stir the proton spins over the wide range of frequencies in the broad line proton NMR spectrum.

In the case just discussed, that of the reducing the ^{13}C linewidth by high power irradiation of the proton spectrum, the experiment is straightforward, and limited primarily by power considerations. This heteronuclear decoupling differs from homonuclear decoupling in which the linewidth, for example, of the proton resonance is to be reduced by suppressing the dipole–dipole interactions among the protons themselves. In this latter case, the high-powered decoupling irradiation would mask the resonance experiment itself.

In such cases, other approaches have been developed. Several types of pulse sequences have been developed to achieve decoupling among like spins themselves, such as WaHuHa. These are discussed in some detail in Section III.C.

II.C. MAGIC ANGLE SAMPLE SPINNING

As discussed in Section II.A, the rapid and random rotation of molecules in the liquid state tends to average out the line broadening due to the dipole–dipole interaction between nuclear spins, making possible the observation of high resolution features in the NMR spectrum. There is an experimental technique which, under many circumstances, has a similar effect for solid samples: magic angle spinning (MAS). We will show in detail in Chapter IV that not only the dipole–dipole interaction, but also other sources of line broadening such as CSA, have an interaction containing a factor $(3 \cos^2 \theta - 1)$, where θ is the angle between the external magnetic field and the axis about which the sample is caused to spin (see Fig. II.C.1). It is easy to see that under these spinning conditions, if

$$\theta_M = \cos^{-1}(1/3)^{1/2} \tag{II.C.1}$$

then for that particular angle the interaction averages to zero. This angle, θ_M, is called the "magic angle" for the dramatic effect of spinning the sample

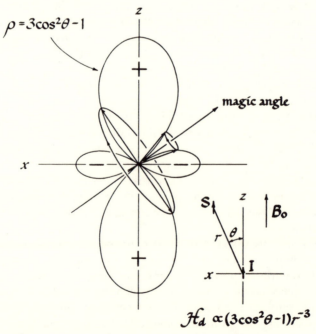

Figure II.C.1   Angular dependence of the dipolar interaction.

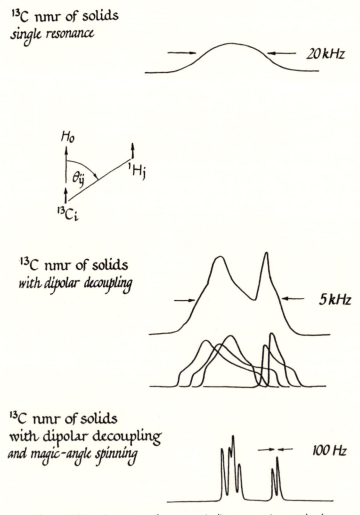

Figure II.C.2   Summary of sparse-spin line narrowing methods.

about the axis inclined at that angle with the direction of the external magnetic field. Note that θ_M is half the tetrahedral angle familiar to organic chemists. The effect of spinning the sample is analogous to the effect of rapid and random molecular rotation in a liquid sample.

Fig. II.C.2 summarizes the conventional line narrowing methods used in sparse-spin NMR.

Other experimental techniques involving somewhat different mechanical manipulations of the sample to reduce line broadening will also be discussed in Section IV: magic angle hopping, double axis rotation, and double rotation. Advantages and disadvantages of these methods are discussed at that point.

II.D. CROSS POLARIZATION

Cross Polarization (CP) is an experimental technique for increasing the sensitivity in the NMR of sparse nuclear spins (such as natural abundance ^{13}C) by making use of those spins' interactions with a collection of abundant spins (such as protons). This sensitivity enhancement arises through two mechanisms. First, one increases the population difference between the spin up and spin down states for the sparse species. Second, one cuts down the time between successive free induction decays (FID) by effectively shortening the relaxation time of the sparse nuclei. In this introductory section, we will give a qualitative description of the technique, aimed at providing a physical feeling for what is going on. Afterward, we will give a semi-rigorous mathematical derivation of the important aspects of the method.

II.D.1. Introduction

To begin with, one prepares the abundant spin system in a state with an artificially low temperature. One then allows the sparse spin system to come into thermal contact with the cold system of abundant spins. Heat flows from the sparse spin system to the cold abundant spin system, which produces a large drop in the spin temperature of the sparse spin system (it is clear that the heat capacity of the abundant spin system is much greater than the heat capacity of the sparse spin system). This accomplishes the first of the two goals mentioned in the introductory paragraph: another way of saying that the spin temperature of the sparse spin system has dropped is to say that the population difference between the upper and lower energy spin states has increased, therefore leading to an increased sensitivity of the NMR experiment.

The second criterion is also met by this method. After a typical NMR FID, one must wait to make another run for a length of time required for the system to reestablish its thermal equilibrium state. Since in a typical experiment one makes many runs and then uses signal averaging to increase the signal-to-noise ratio (see Chapter V), if the time between successive runs can be decreased, more runs can be made in a given period of time. The effect of the good thermal contact between the sparse and abundant spin systems has the effect of reducing the time required for the sparse spin system to reestablish its polarization to the point at which another FID can be obtained.

The first part of the method, putting the abundant spin system into a low spin temperature state, is accomplished by using a $\pi/2$ pulse at the resonance frequency of the abundant nuclei, followed by a spin-locking field in a way that we will describe in detail later. The second essential element of the method, providing good thermal contact between the two spin systems, is tricky. The physical interaction that provides thermal contact between the sparse and abundant spin systems is, in solids, primarily the magnetic dipole–dipole interaction. As we described earlier (see Section IC), such interactions, when time-dependent, give rise to transition probabilities between the Zeeman energy levels, thus altering the population difference between the spin up and spin down states. From a semiclassical standpoint, one can have a resonance exchange of energy between two connected oscillating systems if the natural frequencies of the two systems are close (two pendulums weakly connected to each other undergo a process whereby the energy of oscillation passes back and forth from one pendulum to the other, if the natural frequencies of the two pendulums are near one another). From a quantum mechanical standpoint, exchange of energy is much more likely to occur if the time-dependence of the coupling has a frequency matching the Bohr condition for the difference between energy states. In both approaches, if the natural frequencies of the two systems can be brought near one another, the transition probability for an exchange of energy, and hence the thermal contact between the two systems, is substantially increased.

Since the magnetogyric ratios of the sparse and abundant spins are typically quite different (γ_C is approximately equal to one-fourth of γ_H), then the Larmor frequencies are quite different, and an exchange of energy between the two spin systems is not very likely. Cross polarization depends upon an ingenious method for matching frequencies between the two spin systems using double resonance techniques, to be described shortly.

II.D.2. Cooling the Abundant Spin System

We now address the question of preparing the abundant spin system in a low spin temperature state. One begins with a brief, strong $\pi/2$ pulse of radiation at the resonance frequency which rotates the abundant spin magnetization into the xy plane. If the $\pi/2$ pulse was applied along the x-axis, the magnetization will be rotated from the z-axis to the y-axis in the rotating frame. As described in Section I.B.5, the spin-locking procedure then uses continuous irradiation along the y-axis, so that the magnetization will precess about B_{eff} in the rotating frame.

The magnetization in the rotating frame will decay with a time constant $T_{1\rho}$. Note that in the precession about B_{eff}, the spin temperature of the abundant spin magnetization is very low. Physically, this is because the magnetization is proportional to the excess number of spins in the lower energy state in the strong external magnetic field B_0:

$$N(-1/2)/N(+1/2) = \exp(-\hbar\gamma_I B_0/kT_L), \qquad \text{(II.D.1)}$$

where T_L is the temperature of the lattice (see Section I.A.4).

In the spin-locked state in the rotating frame, the excess number of spins in the lower energy state is given by

$$N(-1/2)/N(+1/2) = \exp(-\hbar\gamma_I B_{\text{eff}}/kT_S) \qquad \text{(II.D.2)}$$

where T_S is the spin temperature.

Comparing Eqs. II.D.1 and II.D.2 we see that the temperature of the abundant spin system is related to the temperature of the lattice by

$$T_S/T_L = B_{\text{eff}}/B_0. \tag{II.D.3}$$

Since $B_{\text{eff}} \ll B_0$, this means that $T_S \ll T_L$.

We have then, a large magnetization for the abundant spin system spin-locked on the y-axis in the rotating frame, precessing about the y-axis with an angular frequency ω_{II}, with the spin temperature much lower than the temperature of the lattice.

At the same time the spin-locking field is used for the abundant spin system, one also subjects the sparse spin system to rf radiation at its own resonance frequency in the external magnetic field B_0:

$$\omega_0 = \gamma_S B_0. \tag{II.D.4}$$

This rf field is continuous (see Fig. II.D.1), so that in the rotating frame the sparse spin magnetization precesses about B_{1S} with an angular frequency given by

$$\omega_{1S} = \gamma_S B_{1S}. \tag{II.D.5}$$

To achieve this, of course, we need a double resonance probe.

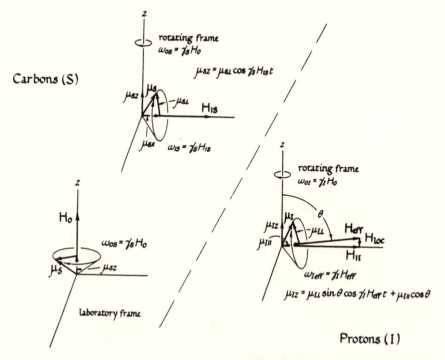

Figure II.D.1 Communication between two dissimilar spin systems; the carbon system is shown both with and without rf irradiation; the proton system is irradiated in the presence of a small local field. Reprinted by permission from John Wiley & Sons, Inc., *Topics in Carbon-13 NMR Spectroscopy*, **3**, Ed. by Levy, Copyright © 1979, Schaefer and Stejskal.

II.D.3. The Hartmann–Hahn Condition

Note that from Fig. II.D.1 both the abundant spin magnetization and the sparse spin magnetization have components along the z-axis. If the angular frequencies ω_{1I} and ω_{1S} are adjusted by varying the oscillating fields amplitudes to satisfy

$$\gamma_S B_{1S} = \gamma_I B_{1I}, \tag{II.D.6}$$

so that

$$\omega_{1S} = \omega_{1I}, \tag{II.D.7}$$

then the z components of the magnetizations for the abundant system and the sparse system have the same time-dependence, and a resonance exchange of energy between the two can take place readily through a mutual spin flip mechanism. The frequency condition given in Eq. II.D.6 is called the Hartmann–Hahn condition, in honor of their observation that such an experimental adjustment can lead to good thermal contact between the sparse and abundant spin systems. The mechanism of thermal contact between the two spin systems (made effective by experimentally satisfying the Hartmann–Hahn condition) will lead to a warming up (slight, because of its large heat capacity) of the abundant spin system, and a considerable cooling down of the sparse spin system. This cooling down is manifested by a significant decrease in the spin temperature of the sparse spin system down to

$$T_S^{cp} = T_I = (B_{1I}/B_0)T_L = (\gamma_S/\gamma_I)(B_{1S}/B_0)T_L, \tag{II.D.8}$$

where we have used Eqs. II.D.3 and II.D.6. Since the spin temperature of the S spin that would have been achieved by direct spin locking of the S spin is

$$T_S^0 = (B_{1S}/B_0)T_L, \tag{II.D.9}$$

we have lowered T_S by a factor γ_S/γ_I. For protons and ^{13}C nuclei, this is a factor of $1/4$. This means, through Eq. II.D.1, an increased sensitivity in a free induction decay resonance experiment of the sparse spin system. Note that when the cross polarization experiment begins there is no S spin polarization along the rf field. This corresponds to infinite T_S. Cross polarization reduces the temperature all the way to T_S^{cp}, Eq. II.D.8.

To summarize the experiment, one applies a $\pi/2$ pulse to the abundant spin system, followed by a change of phase leading to spin-locking of the abundant spin system (see Fig. II.D.2). The sparse spins are also spin locked but without significant polarization, and the precession of the I and S magnetizations are adjusted to satisfy the Hartmann–Hahn condition, and the systems are kept in thermal contact for a while. One follows the contact period by an FID observation of the sparse spin system (see Fig. II.D.3).

II.D.4. Descriptions of CP Using the Solomon Equations

To gain further insight into the cross polarization process, we will examine it in the light of the predictions of the Solomon equations for the rotating frame.

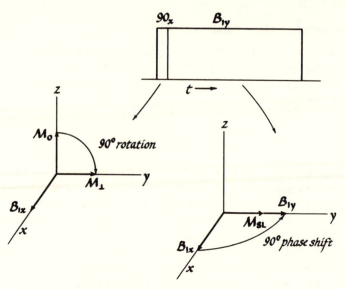

Figure II.D.2   Spin locking the protons.

Cross Polarization

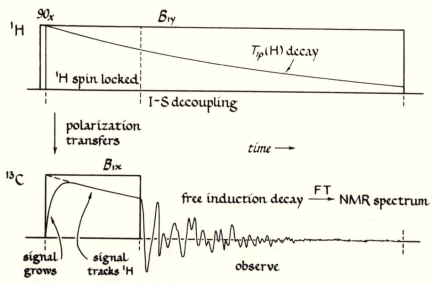

Figure II.D.3   Hartmann–Hahn single-contact, cross-polarization sequence.

Consider the case of many interacting spins, working in analogy with the two spin case. For the interaction of two spins i and j that are both in the I system, we have

$$-(dI_i/dt)_j = \rho_{Iji}I_i + \sigma_{Iij}I_j + \text{const.} \tag{II.D.10}$$

and

$$-(dI_j/dt)_i = \rho_{Iji}I_j + \sigma_{Iji}I_i + \text{const.} \tag{II.D.11}$$

For the interaction between two spins i and k, with i in the I spin system and k in the S spin system, we have

$$-(dI_i/dt)_k = \rho_{ki}I_i + \sigma_{ik}S_k + \text{const.} \tag{II.D.12}$$

and

$$-(dS_k/dt)_i = \rho'_{ik}S_k + \sigma_{ki}I_i + \text{const.}, \tag{II.D.13}$$

where the parameters ρ_1, ρ'_1 and σ are defined in analogy with Eqs. I.D.16–I.D.18; clearly, $\rho_{Iij} = \rho_{Iji}$, $\sigma_{Iji} = \sigma_{Iij}$, $\sigma_{ki} = \sigma_{ik}$, but ρ'_{ik} is not, in general, equal to ρ_{ki}.

We now have

$$-dI/dt = -\sum_i dI_i/dt = -\sum_i \left[\sum_j (dI_i/dt)_j + \sum_k (dI_i/dt)_k \right] \tag{II.D.14}$$

where the summation goes from 1 to n_I for the I spins and from 1 to n_S for the S spins.

We will assume that the spins are abundant and the S spins are sparse; for example, protons could be the I spins and ^{13}C could be the S spins in a typical experiment. Since the S spins are sparse, we will not consider pairwise interactions among them.

If we write

$$I = \sum_i I_i \quad \text{and} \quad S = \sum_k S_k, \tag{II.D.15}$$

we have

$$-dI/dt = \sum_i I_i \left[\sum_j (\rho_{Iji} + \sigma_{Iji}) + \sum_k \rho_{ki} \right] + \sum_k S_k \sum_i \sigma_{ik} + \text{const.}, \tag{II.D.16}$$

where we have set $\rho_{Iii} = \sigma_{Iii} = 0$ owing to the fact that a spin will not interact with itself, and where we have interchanged the order of three summations and one pair of summation indices.

We now write averages of the parameters as

$$\langle (\rho_I + \sigma_I)_i \rangle_{\text{av}} = \sum_j (\rho_{Iji} + \sigma_{Iji})/n_I, \tag{II.D.17}$$

$$\langle (\rho)_i \rangle_{\text{av}} = \sum_k \rho_{ki}/n_S, \tag{II.D.18}$$

and

$$\langle (\sigma)_k \rangle_{\text{av}} = \sum_i \sigma_{ik}/n_I, \tag{II.D.19}$$

so that Eq. II.D.16 becomes

$$-dI/dt = (n_I \langle \rho_I + \sigma_I \rangle_{\text{av}} + n_S \langle \rho \rangle_{\text{av}})I + n_I \langle \sigma \rangle_{\text{av}} S + \text{const.}, \tag{II.D.20}$$

and

$$-dS/dt = n_S \langle \rho_S + \sigma_S \rangle_{\text{av}} + n_I \langle \rho' \rangle_{\text{av}} + n_S \langle \sigma \rangle_{\text{av}} + \text{const.}, \tag{II.D.21}$$

after subscripts have been dropped.

In the usual treatment of the Solomon equations, the constants are evaluated, as before, by requiring that the magnetizations have their thermal equilibrium values in the steady state ($I = I_0$ and $S = S_0$ when $dI/dt = dS/dt = 0$). We will impose, however, different conditions in order to describe cross polarization (CP). In this case, we have both I and S spin-locked in the rotating frame (see Section I.B.5), which corresponds to $I = S = 0$ when $dI/dt = dS/dt = 0$, so that both constants are zero.

We have, under these conditions,

$$-dI/dt = (k_I + f_S k_C)I - f_I k_C S \qquad \text{(II.D.22)}$$

and

$$-dS/dt = (k_S + f_I k_C)S - f_S k_C I, \qquad \text{(II.D.23)}$$

where we have written

$$k_I = 2n_I(\langle W_{I1}\rangle_{\text{av}} + \langle W_{I2}\rangle_{\text{av}}) + 2n_S(\langle W_1\rangle_{\text{av}} + \langle W_2\rangle_{\text{av}}), \qquad \text{(II.D.24)}$$

$$k_S = 2n_S(\langle W_{S1}\rangle_{\text{av}} + \langle W_{S2}\rangle_{\text{av}}) + 2n_I(\langle W_i'\rangle_{\text{av}} + \langle W_2\rangle_{\text{av}}), \qquad \text{(II.D.25)}$$

$$k_C = (n_I + n_S)(\langle W_0\rangle_{\text{av}} - \langle W_2\rangle_{\text{av}}), \qquad \text{(II.D.26)}$$

$$f_I = n_I/(n_I + n_S), \qquad \text{(II.D.27)}$$

and

$$f_S = n_S/(n_I + n_S). \qquad \text{(II.D.28)}$$

K_I and K_S are measures of the thermal contact within and relaxation of the I and S spin systems, and k_C is a measure of the thermal contact between the two systems. K_I, K_S, and k_C depend on the number of spins in each of the systems and on the transition probabilities between the energy levels of the spin systems. The magnitudes of these transition probabilities can be calculated using time-dependent perturbation theory or density matrix theory, with the dipole–dipole interaction between the spins being the time-dependent perturbation inducing the transitions. Those probabilities depend, among other things, on the relationship between the energy difference between spin states: it is here that the Hartmann–Hahn condition plays its key role. By matching energy differences between the Zeeman levels in the two rotating frames, the thermal contact between the two spin systems through k_C is greatly increased.

As can be established by substitution, Eqs II.D.10 and II.D.11 are solved by

$$I = a(+) \exp[r(+)t] + a(-) \exp[r(-)t] \qquad \text{(II.D.29)}$$

and

$$S = b(+) \exp[r(+)t] + b(-) \exp[r(-)t], \qquad \text{(II.D.30)}$$

where

$$r(\pm) = (1/2)[-(K_I + K_S + k_C) \pm R] \qquad \text{(II.D.31)}$$

and

$$R = |\{[K_I + K_S) + (f_S - f_I)k_C]^2 + 4f_I f_S k_C^2\}^{1/2}|. \qquad \text{(II.D.32)}$$

We determine $a(\pm)$ and $b(\pm)$ for the CP case by using the following initial conditions: at $t = 0$, $I = I_0$ (where I_0 is the initial polarization in the rotating frame after the $\pi/2$ pulse, and $S = 0$. This gives

$$b(+) = -b(-) = A \qquad \text{(II.D.33)}$$

and

$$a(+) + a(-) = I_0. \qquad \text{(II.D.34)}$$

We now use the original equation

$$-dI/dt = (K_I + f_S k_C)I - f_I k_C S$$
$$= -a(+)r(+)\exp[r(+)t] - a(-)r(-)\exp[r(-)t], \qquad \text{(II.D.35)}$$

where the right-hand side follows from differentiating Eq. II.D.29.

Evaluating this at $t = 0$ (where $I = I_0$ and $S = 0$) and solving simultaneously with Eqs. II.D.31 and II.D.34 leads to

$$a(\pm) = (I_0/2)[1 \mp (K_I - K_S + \{f_S - f_I\}k_C/R]. \qquad \text{(II.D.36)}$$

We now consider some special cases in order to put some physical content into the mathematical expressions.

a. If there is no S system, Eq. II.D.35 reduces to

$$-(dI/dt) = K_I I \qquad \text{(II.D.37)}$$

which has the solution

$$I = I_0 \exp(-K_I t) = I_0 \exp(-t/T_{1\rho}), \qquad \text{(II.D.38)}$$

where we have written

$$T_{1\rho} = 1/K_I \qquad \text{(II.D.39)}$$

for the longitudinal relaxation time in the rotating frame.

b. Now consider the case in which there is no relaxation except through the contact between the I and S systems, that is, assume K_I and K_S are zero. We have, then,

$$R = |[(f_S - f_I)^2 k_C^2 + 4f_I f_S k_C^2]^{1/2}|$$
$$= (f_I + f_S)k_C = k_C, \qquad \text{(II.D.40)}$$

where we have used Eq. II.D.28.

Then

$$r(\pm) = (1/2)(-k_C \pm k_C), \qquad \text{(II.D.41)}$$

or

$$r(+) = 0 \qquad \text{and} \qquad r(-) = -k_C. \qquad \text{(II.D.42)}$$

Also, if $K_I = K_S = 0$, we have, from Eq. II.D.36,

$$a(\pm) = (I_0/2)[1 + (f_S - f_I)k_C/k_C] \qquad \text{(II.D.43)}$$

or

$$a(+) = f_I I_0, \qquad a(-) = f_S I_0, \qquad \text{(II.D.44)}$$

since $f_S + f_I = 1$.

Finally,

$$I = I_0[f_I + f_S \exp(-k_C t)] \qquad (\text{II.D.45})$$

and

$$S = A[1 - \exp(-k_C t)]. \qquad (\text{II.D.46})$$

The important thing to note is that the sparse spin polarization starts at zero at $t = 0$, and then increases, as the negative term decreases, at a rate that is large when k_C (a measure of the thermal contact between I and S systems) is large.

c. If the S system is very small relative to the I system, we have $n_S \ll n_I$, so that

$$f_S/f_I \ll 1 \qquad \text{and} \qquad f_I \approx 1. \qquad (\text{II.D.47})$$

The expression for R in Eq. II.D.32 reduces, in this limit, to

$$R = |\{[(K_I - K_S) + (f_S - f_I)k_C]^2 + 4f_I f_S k_C^2\}^{1/2}| \approx |K_I - K_S - k_C|; \qquad (\text{II.D.48})$$

furthermore, since communication between the systems is usually more efficient that spin-lattice relaxation, then, typically, k_C is greater than K_I and K_S, so that $k_C + K_S > K_I$. Using the fact that all are positive, we have

$$R = |K_I - K_S - k_C| = k_C + K_S - K_I. \qquad (\text{II.D.49})$$

Then

$$r(\pm) = 1/2[-(K_I + K_S + k_C) \pm (k_C + K_S - K_I)], \qquad (\text{II.D.50})$$

so that

$$r(+) = -K_I = -1/T_{1\rho} \qquad (\text{II.D.51})$$

and

$$r(-) = -(K_S + k_C). \qquad (\text{II.D.52})$$

Under these conditions, Eq. II.D.36 becomes

$$a(\pm) = (I_0/2)[1 \mp (K_I + K_S - k_C)/(k_C + K_S - K_I)], \qquad (\text{II.D.53})$$

so that

$$a(+) = I_0, \qquad a(-) = 0. \qquad (\text{II.D.54})$$

Finally, we have, for the case of abundant I and sparse S,

$$I = I_0 \exp(-K_I t) \qquad (\text{II.D.55})$$

and

$$S = A\{\exp(-K_I t) - \exp(-[K_S + k_C]t)\}. \qquad (\text{II.D.56})$$

From the form of this equation, one can make the identifications of $T_{1\rho}$ with K_I^{-1}, and T_{IS}, with k_C^{-1}. Although K_S, a dissipative relaxation appears to assist k_C, we shall see, later, that it has the effect of reducing S (Eq. II.D.70).

The abundant spin magnetization in the rotating frame, that is, starts out at I_0 after the $\pi/2$ pulse, and decreases exponentially to zero with the characteristic time constant $T_{1\rho}$. The sparse spin magnetization starts at zero, increases through its interaction with the abundant spins, and then decays to zero (see Fig. II.D.3).

II.D.5. Spin Temperature Equilibration

Next, let us examine quantitatively the polarization enhancement of the sparse spin system due to the CP process.

We will first calculate the spin temperature of the abundant spin system immediately before and after the $\pi/2$ pulse. For a spin 1/2 system, a population ratio of the down and up spin states given by the Boltzmann equilibrium distribution:

$$N(-1/2)/N(+1/2) = \exp(-\hbar\omega/kT_S) \approx 1 - \hbar\omega/kT_S, \quad \text{(II.D.57)}$$

where we have used the high temperature approximation (well satisfied for NMR frequencies); T_S is the spin temperature. We will write

$$\beta_S = \hbar/kT_S, \quad \text{(II.D.58)}$$

$$\Delta N = N(+1/2) - N(-1/2), \quad \text{(II.D.59)}$$

and

$$N(+1/2) \approx N(-1/2) \approx N/2 \quad \text{(II.D.60)}$$

where N is the total number of spins. It follows quickly that

$$\Delta N \approx N\omega\beta_S/2. \quad \text{(II.D.61)}$$

For the I spin system just before the $\pi/2$ pulse, we have

$$\Delta N = N\omega_0\beta_L/2 \quad \text{(II.D.62)}$$

where the subscript L refers to the temperature of the lattice in which the spin systems find themselves. After the $\pi/2$ pulse and the spin lock,

$$\Delta N = N\omega_{1I}\beta_I/2, \quad \text{(II.D.63)}$$

so that

$$\beta_I/\beta_L = T_L/T_I = \omega_0/\omega_{1I} = B_0/B_{1I}. \quad \text{(II.D.64)}$$

The I spin system is now at a very cold temperature relative to T_L, by a ratio of the strong external field to the magnitude of the spin-lock field, which is several orders of magnitude.

When the S spin system is spin locked by B_{1S}, we have

$$\beta_S = (B_0/B_{1S})\beta_L. \quad \text{(II.D.65)}$$

Now the I and S spin systems are brought into good thermal contact by satisfying the Hartmann–Hahn condition:

$$\gamma_S B_{1S} = \gamma_I B_{1I}. \quad \text{(II.D.66)}$$

If we neglect relaxation, the two spin systems will equilibrate, so that

$$\beta'_S = \beta_I = (B_0\beta_L)/B_{1I} = (\gamma_I/\gamma_S)(B_0/B_{1S})\beta_L, \qquad \text{(II.D.67)}$$

or

$$\beta'_S = (\gamma_I/\gamma_S)\beta_S, \qquad \text{(II.D.68)}$$

where we have assumed that the spin temperature of the abundant spin system does not change during the equilibration since $N_I \gg N_S$. Since ΔN is proportional to β, the CP process enhances the sensitivity of a measurement of S by a factor

$$\varepsilon = \gamma_I/\gamma_S. \qquad \text{(II.D.69)}$$

If the I spins are protons and the S spins ^{13}C nuclei, ε is about four.

We next return to Eq. II.D.23. Using $f_I \approx 1$, differentiating Eq. II.D.56 for the left-hand side of Eq. II.D.23 and using Eqs. II.D.55 and II.D.56 for the right-hand side, and setting $t = 0$, we obtain

$$A = f_S k_C I_0/(K_S + k_C - K_I), \qquad \text{(II.D.70)}$$

so that

$$S = [f_S k_C I_0/(K_S + k_C - K_I)][\exp(-K_I t) - \exp(-\{K_S + k_C\}t)]. \quad \text{(II.D.71)}$$

If, as is sometimes the case, $k_C = K_S < K_I$, this equation is still valid, but the roles of K_I and $k_C + K_S$ are interchanged.

Recall that in the equilibrium state (in the absence of CP), the ratio of the polarization of sparse spins to that of the abundant spins can be written

$$S_0/I_0 = (\gamma_S/\gamma_I)(f_S/f_I) \approx (\gamma_S/\gamma_I)f_S, \qquad \text{(II.D.72)}$$

since we are taking $f_I \approx 1$.

Solving this for I_0, substitution into Eq. II.D.71, taking $k_C \gg K_I, K_S$, and doing some algebra, we obtain

$$S = S_0(\gamma_I/\gamma_S)[\exp(-K_I t) - \exp(-k_C t)]. \qquad \text{(II.D.73)}$$

Since $k_C > K_I$, the sparse spin polarization will increase from zero, and then decay with a time constant approximating that of the abundant spin system. The quantity $S_0(\gamma_I/\gamma_S)$ can be defined to be S_E, the "enhanced" polarization arising from the CP procedure.

In this section, we made assumptions about the relative sizes of K_I, K_S, and k_C in order to obtain an expression for a typical CP evolution. In many systems a different hierarchy may be found, the result of which is easily derived from the general equations. The results are similar to those presented.

The CP experiment was originally designed to detect the sparse spins by observing the effect on the abundant spin resonance after multiple contacts with the sparse spin system. When the Hartmann–Hahn condition is satisfied, there will be maximal effect on the I spin resonance due to transfer to polarization to the S spin system.

We have described the CP experiment illustrated in Fig. II.D.4 by observing a single FID of the S system after cross polarization takes place. One may also obtain multiple FID's following multiple contacts between

matched CP

single contact

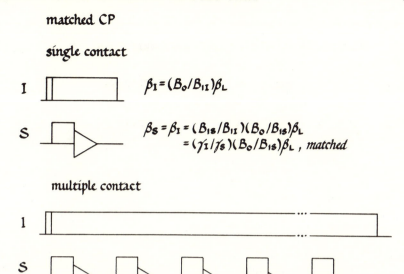

$$\beta_I = (B_o/B_{1I})\beta_L$$

$$\beta_S = \beta_I = (B_{1S}/B_{1I})(B_o/B_{1S})\beta_L$$
$$= (\gamma_I/\gamma_S)(B_o/B_{1S})\beta_L \, , \, matched$$

multiple contact

Figure II.D.4 Single- and multiple-contact, matched cross-polarization sequences.

the S and I spin system during a single spin lock of the abundant spin system if its $T_{1\rho}$ is long enough, and use signal averaging to improve S/N ratio (see Fig. II.D.4).

A different experimental technique used to achieve a cool spin temperature for the I spin system is called adiabatic demagnetization in the rotating frame (ADRF). One first brings the I polarization to the y-axis of the rotating frame by a $\pi/2$ pulse. One then starts with a field B_{1y} constant in the rotating frame, as in the case of spin-locking, but then reduces B_{1y} adiabatically (reducing it to zero in a time long compared with the inverse line width of the I signal), so that the spins are left in the local dipolar field due to neighboring spins. The result is an even lower proton spin temperature but also a reduced coupling between the spin systems. Coupling requires that the proton fluctuation spectrum, a broad line centered about $\omega_{1I} = 0$ for the I spins shown in Fig. II.D.5, have a tail which overlaps the resonance of the S spin system. One advantage of this experiment is that, since the Hartmann–Hahn condition need not be satisfied, one is free to use a large B_{1S}, which, through Eq. II.D.2, increases the polarization of the S spins. The disadvantage is that thermal contact between the I and S systems is much weaker the larger B_{1S} gets, so that transfer of polarization from the I to the S system is much slower.

II.D.6. Calculation of k_C

We now outline the steps involved in calculating k_C from first principles following. In presenting the final form, we will indicate that satisfaction of the Hartmann–Hahn condition is a critical component in determining how fast the polarization transfer takes place. Following traditional notation,

Adiabatic Demagnetization in the Rotating Frame

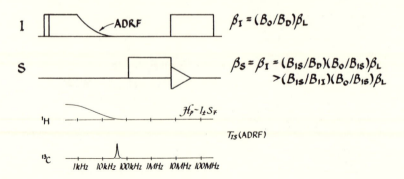

Figure II.D.5 ADRF sequence. Reprinted by permission from John Wiley & Sons, Inc., *Topics in Carbon-13 NMR Spectroscopy*, **3**, Ed. by Levy, copyright © 1979, Schaefer and Stejskal.

we will write k_C as an inverse relaxation time:

$$k_C = T_{IS}^{-1}. \tag{II.D.74}$$

One begins by writing the Hamiltonian for the whole spin system as a sum of three terms: one for the I spins, one for the S spins, and one representing the interaction between the two, which depends on coordinates of both I and S spins. The interaction term is then treated as a time-dependent perturbation on the energy levels determined for the other two terms. The calculation is similar to that given in Section I.D.3 for the theory of relaxation times.

The specific form of the interaction term in the laboratory frame is (see Section III.A; this is the secular part of the dipole–dipole Hamiltonian):

$$H_{IS} = 2\gamma_I\gamma_S k^2 \sum_{i=1}^{N_I} \sum_{j=1}^{N_I} r_{ij}^{-3}(3\cos^2\theta_{ij} - 1)I_{iz}S_{jz}/2, \tag{II.D.75}$$

where r_{ij} and θ_{ij} are the polar coordinates of the vector connecting spins I_i and S_j. One transforms to the double rotating frame and then calculates a transition probability per unit time due to the perturbation. T_{IS} will be inversely proportional to the transition probability: the greater that probability, the faster polarization is transferred. One obtains, to a good approximation,

$$(T_{IS})^{-1} = (\pi^{1/2}/4)\sin^2\theta_S \sin^2\theta_I M_2^{IS}\tau_C \exp[-(\omega_{1S} - \omega_{1I})^2\tau_C^2/4], \tag{II.D.76}$$

where M_2^{IS} is the second moment of the dipolar coupling between an I and an S spin, where θ_S and θ_I are defined by

$$\tan\theta_I = \omega_{1I}/(\omega_0 - \omega_{1I}) \tag{II.D.77}$$

and

$$\tan\theta_S = \omega_{1S}/(\omega_0 - \omega_{1S}), \tag{II.D.78}$$

and where τ_C is the correlation time, defined analogously to that encountered in Section I.D.3 in the discussion of relaxation. It is a measure of the interaction time of the two spins. One influence that will limit τ_C is the lifetime of the I spin states as revealed by the I–I fluctuation spectrum.

Note that the gaussian factor in the expression for $(T_{IS})^{-1}$ has its maximum for frequencies satisfying the Hartmann–Hahn condition. The shorter τ_C is the wider will be the gaussian and the less critical the Hartmann–Hahn match.

II.D.7. CP Combined with MAS

When CP is combined with MAS (magic angle spinning, as discussed in Section II.C), there is a possible complication to which the experimenter should be alert. This arises because sample spinning can give an oscillatory time-dependence to the dipolar interaction between I and S spins that provides the thermal contact leading to polarization transfer. This can affect the CP process if the sample spinning rate is of the same scale as the rate of the abundant spin dipolar fluctuations: energy-conserving mutual spin slips that give an appreciable line width to the abundant spin resonance (appreciable in comparison with the sparse spin line width, which is small because of abundant spin decoupling and the fact that the sparse spins are far apart).

The crucial parameter is the ratio of the abundant spin line width to the spinning frequency. When this ratio is large (>0.5), which corresponds physically to low spinning speeds, the spinning sidebands (see Section IV.C.1) lie mostly within the proton line width. If the ratio is 0.5 or less, the sidebands, whose spacing is just the spinning frequency, fall outside the proton line width. This can lead to a modification of the CP polarization transfer. The most graphic way to examine this effect is to generate what is called a Hartmann–Hahn spectrum, a plot of $(T_{IS})^{-1}$ (or signal intensity for a fixed contact time) as a function of either ω_{1I} or ω_{1S} with the other held constant (Fig. II.D.6).

One can measure the CP transfer rates by doing a series of experiments in which the contact time (see Fig. II.D.3) between the I and S spins is varied. Fig. II.D.7 presents the results of such an experiment on poly(methyl methacrylate), and it is clear that the transfer rates can differ from one site to another. The carbonyl carbon line, appearing on the left side of the spectrum, builds up more slowly than the line due to protonated carbons. This is reasonable because the proximity of a proton enhances the dipolar interaction, which, in turn, facilitates polarization transfer.

II.E. MEASURING $T_{1\rho}$

The spin lattice relaxation time for the abundant spins in the rotating frame, written $T_{1\rho}$ (or $T_{1\rho}(I)$ to contrast with $T_{1\rho}(S)$) describes the rate at which a nuclear magnetization, spin-locked in the rotating frame, reapproaches its thermal equilibrium value. In the spin-locking process, a magnetization

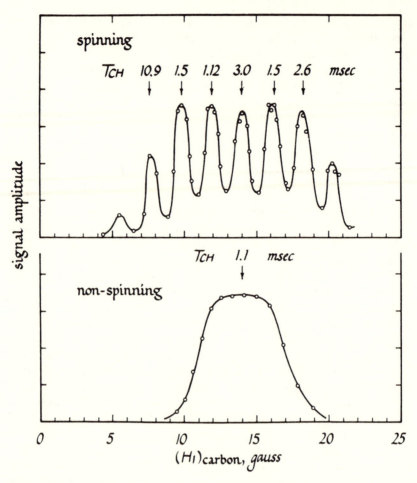

Figure II.D.6 Hartmann–Hahn spectra showing the effect of MAS on the CP transfer process. Reprinted by permission from Academic Press Permissions, *J. Magn. Reson.*, **28** (1977), Stejskal, Schaefer, and Waugh.

correspond to the strong external magnetic field B_0 is brought into the x–y plane by a $\pi/2$ pulse, after which the phase of the radio frequency signal is changed to spin-lock the magnetization on the y axis in the rotating frame. Since the effective magnetic field in the rotating frame, B_{eff}, is orders of magnitude smaller than B_0, the rotating frame magnetization is much larger than its thermal equilibrium value. It will decay toward that lower value with the time constant $T_{1\rho}$.

$T_{1\rho}$ can be measured directly by a series of experiments in each of which the spin-lock field is maintained for a particular length of time, which is to be varied from run to run of the experiment; when the spin-lock field is removed, one obtains FID whose amplitude decreases with increasing spin-lock time, as the magnetization decays (see Fig. II.D.3). A semilog plot of spin-locked time and FID amplitude allows $T_{1\rho}$ to be inferred. $T_{1\rho}$ can

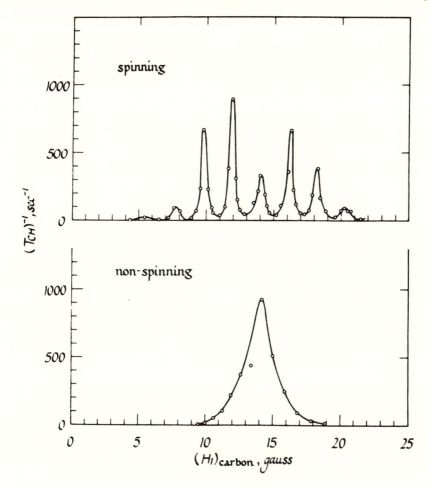

Figure II.D.6 (continued).

be measured indirectly in the CP experiment by a series of experiments in which the contact time is increased incrementally. Since the rare spin polarization tracks the $T_{1\rho}$ decay, the rare spin spectrum can be used to determine $T_{1\rho}$ (see Fig. II.D.7).

Note that if there were no spin-lock field after the original $\pi/2$ pulse bringing the x–y magnetization into the plane, the relaxation time would simply be T_2. On the other hand, if B_{eff} equalled B_0, any non-equilibrium magnetization would decay with time constant T_1. Since $T_2 \leq T_1$, we have the relationship

$$T_2 \leq T_{1\rho} \leq T_1. \tag{II.E.1}$$

In a system which is not heterogeneous on a macroscopic scale, mutual spin flips between protons maintain a uniform proton spin temperature, even though there may be different relaxation rates from different protons. $T_{1\rho}$ depends on lattice motions to induce transitions through the proton–proton

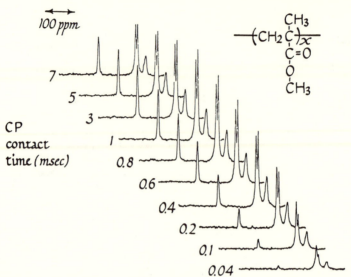

Figure II.D.7 CP spectra for varying contact times. Reprinted with permission from the American Chemical Society, *Macromolecules*, **10**, © 1977, Schaefer, Stejskal, and Buchdahl.

dipolar coupling. Fluctuating dipole–dipole interactions correspond to fluctuating magnetic fields which act like electromagnetic radiation inducing transitions between different spin energy levels. Random spin flips in which there is no net change in magnetization are responsible for cross polarization and spin diffusion. $T_{1\rho}$ for protons, however, involves paired spin flips and lattice energy.

Note that the I–I spin flips due to the dipole–dipole interaction not only provide line broadening in the I system, but also, through the dipole–dipole interaction between the I and S spins, line broadening in the S spin system as well. This line broadening is homogeneous in the sense described earlier because its effect is to limit the lifetime of the I–S interaction. The abundant proton spin system has a fluctuation spectrum which is related to W_{10} from our discussion of the Solomon Equations. An additional effect of the flip-flop interactions in the abundant spin system is to bring about spin diffusion.

The rare spin S has a $T_{1\rho}$ as well, which we will call $T_{1\rho}(S)$ to contrast it to $T_{1\rho}(I)$ for the abundant spin. $T_{1\rho}(S)$ can be measured directly by spin locking the S spins as we did the I spins. It is more usual to use CP to generate spin locked S polarization and then remove the I irradiation to allow the S spins to relax on their own. For observation purposes, I–S decoupling is restored after S irradiation is turned off (see Fig. II.E.1).

The nature of the mechanisms that determine $T_{1\rho}(I)$ and $T_{1\rho}(S)$ will be discussed in Section III.C.2. These two relaxation times are complementary since they are determined by different mechanisms ($T_{1\rho}(S)$ has two competing

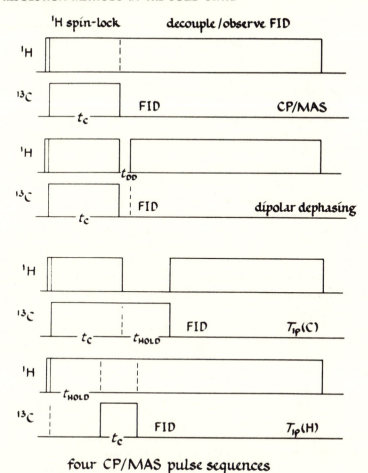

four CP/MAS pulse sequences

Figure II.E.1 CP variations. Reprinted with permission by VCH Publishers, © 1992, *Polymer and Fiber Science: Recent Advances*, edited by Fornes and Gilbert.

mechanisms) and they average system inhomogeneity differently (the S system does not engage in spin diffusion).

Fig. II.E.1 also pictures two other common CP sequences. The preferred way to measure $T_{1\rho}(I)$ indirectly is shown. In this sequence, a CP contact of fixed duration is moved along the $T_{1\rho}(I)$ decay to observe its progress. The sequence separates T_{IS} behavior from $T_{1\rho}(I)$ by ignoring the former. Also, rf heating by the S irradiation is independent of hold time. The other sequence shown in the figure is called dipolar dephasing (DD) or sometimes delayed decoupling or interrupted decoupling. It is a spectrum editing technique used to suppress rare spins strongly coupled to the abundant spin system (say $-CH_2-$ and $-CH\diagup$ signals) because they decay rapidly during the time decoupling is withheld. Other signals ($\diagup C \diagdown$ and $-CH_3$, for instance) survive since their decay is slower. This experiment is closely related to DIPSHIFT and DRSE (Section IV.E.2 and IV.E.3).

II.F. EXPERIMENTAL CONSIDERATIONS

The width of NMR absorption lines in solids, as opposed to liquids, requires high power irradiation. A solid line width of 50 kHz, requiring a $\pi/2$ pulse lasting a few microseconds, will require power in the kW range.

Another important consideration has to do with the relationship between the frequency of magic angle spinning f_s relative to other parameters of the system. The spinning frequency should be greater than the CSA maximum spectral splitting (see Section IV.C.1), which would require it to be of the order of kHz. On the other hand, f_s would be considerably less than the magnitude of the proton dipole–dipole interaction, which is about 30 kHz, and also considerably less than the dipole–dipole interaction between a proton and the ^{13}C nucleus, which is of the order of 20 kHz. In order to avoid the complications described in Section II.D.7, f_s should also differ from the reciprocals of any appropriate correlation times.

Decoupling fields in solid state experiments must also be high power: a typical decoupling field under these circumstances needs to exceed the local field due to the protons (or proton line width), in other words, more than 30 kHz.

It may be helpful to give some numerical ranges for the parameters appearing in a consideration of the CP/MAS experiment in a relatively rigid solid with some molecular motion. For T_{IS} the range varies from about 50 μs for protonated carbons to about 500 μs for quaternary carbons. $T_2^*(C)$, a measure of the decay of ^{13}C FID, varies from 50 μs for protonated carbons to 500 μs for quaternary carbons. $T_2^*(H)$ is typically in the range of 10 to 30 μs. The spin lattice relaxation time in the rotating frame for protons, $T_{1\rho}(H)$, is of the orders 5 to 50 ms, whereas $T_1(H)$ is of the order of a few seconds. The spin lattice relaxation for the carbons $T_1(C)$ may be up to 100 seconds. The root-mean-square line width for the proton signal is of the order of 25 kHz.

BIBLIOGRAPHY

Books

Fyfe, Colin A., *Solid State NMR for Chemists*, C.F.C. Press, Guelph, Ontario, Canada, 1983.

Articles

Chapter II

Etter, M. C., Hoye, R. C., and Vojta, G. M. (1988) "Solid-State NMR and X-Ray Crystallography: Complementary Tools for Structure Determination," *Crystallography Reviews* **1**, 281–338.

II.A

VanderHart, D. L., Earl, W. L., and Garroway, A. N. (1981) "Resolution in ^{13}C NMR of Organic Solids Using High-Power Proton Decoupling and Magic-Angle Sample Spinning," *J. Magn. Reson.* **44**, 361–401.

II.D

Hartmann, S. R. and Hahn, E. L. (1962) "Nuclear Double Resonance in the Rotating Frame," *Phys. Rev.* **128**, 2042–2053.

Lurie, F. M. and Slichter, C. P. (1964) "Spin Temperature in Nuclear Double Resonance,"*Phys. Rev.* **133**, A1108–A1122.

Müller, L., Kumar, A., Baumann, T., and Ernst, R. R. (1974) "Transient Oscillations in NMR Cross-Polarization Experiments in Solids," *Phys. Rev. Lett.* **32**, 1402–1406.

Pines, A., Gibby, M. G., and Waugh, J. S. (1971) "Proton-Enhanced Nuclear Induction Spectroscopy. A Method for High Resolution NMR of Dilute Spins in Solids," *J. Chem. Phys.* **56**, 1776–1777.

Pines, A., Gibby, M. G., and Waugh, J. S. (1973) "Proton-enhanced NMR of Dilute Spins in Solids,"*J. Chem. Phys.* **59**, 569–590.

II.D.4

Bloembergen, N. (1956) "Spin Relaxation Processes in a Two-Proton System," *Phys. Rev.* **104**, 1542–1547.

Brooks, A. A., Cutnell, J. D., Stejskal, E. O., and Weiss, V. W. (1968) "Proton NMR Relaxation Effects. Cross-Relaxation Processes in Pure Liquids,"*J. Chem. Phys.* **49**, 1571–1576.

Naito, A., Ganapathy, S., Akasaka, K., and McDowell, C. A. (1983) "Spin-Lattice Relaxation of ^{13}C in Solid Amino Acids Using the CP-MAS Technique," *J. Magn. Reson.* **54**, 226–235.

II.D.5

Lurie, F. M. and Slichter, C. P. (1964) "Spin Temperature in Nuclear Double Resonance," *Phys. Rev.* **133**, A1108–A1122.

II.D.7

Demco, D. E., Tegenfeldt, J., and Waugh, J. S. (1975) "Dynamics of Cross Relaxation in Nuclear Magnetic Double Resonance," *Phys. Rev. B* **11**, 4133–4151.

Schaefer, J. and Stejskal, E. O. (1976) "Carbon-13 Nuclear Magnetic Resonance of Polymers Spinning at the Magic Angle," *J. Am. Chem. Soc.* **98**, 1031–1032.

Stejskal, E. O., Schaefer, J., and Waugh, J. S. (1977) "Magic-Angle Spinning and Polarization Transfer in Proton-Enhanced NMR," *J. Magn. Reson.* **28**, 105–112.

II.E

Schaefer, J., Sefcik, M. D., Stejskal, E. O., and McKay, R. A. (1981) "Magic-Angle Carbon-13 Nuclear Magnetic Resonance Analysis of the Interface between Phases in a Blend of Polystyrene with a Polystyrene–Polybutadiene Block Copolymer," *Macromolecules* **14**, 188–192.

Schaefer, J., Sefcik, M. D., Stejskal, E. O., McKay, R. A., Dixon, W. T., and Cais, R. E. (1984) "Molecular Motion in Glassy Polystyrenes," *Macromolecules* **17**, 1107–1118.

Schaefer, J., Stejskal, E. O., and Buchdahl, R. (1975) "High-Resolution Carbon-13 Nuclear Magnetic Resonance Study of Some Solid, Glass Polymers," *Macromolecules* **8**, 291–296.

Schaefer, J., Stejskal, E. O., and Buchdahl, R. (1977) "Magic-Angle ^{13}C NMR Analysis of Motion in Solid Glassy Polymers," *Macromolecules* **10**, 384–405.

Schaefer, J., Stejskal, E. O., Steger, T. R., Sefcik, M. D., and McKay, R. A. (1980) "Carbon-13 $T_{1\rho}$ Experiments on Solid Glassy Polymers," *Macromolecules* **13**, 1121–1126.

Sefcik, M. D., Schaefer, J., Stejskal, E. O., and McKay, R. A. (1980) "Analysis of the Room-Temperature Molecular Motions of Poly(ethylene terephthalate)," *Macromolecules* **13**, 1132–1137.

Steger, T. R., Schaefer, J., Stejskal, E. O., and McKay, R. A. (1980) "Molecular Motion in Polycarbonate and Modified Polycarbonates," *Macromolecules* **13**, 1127–1132.

Stejskal, E. O., Schaefer, J., Sefcik, M. D., and McKay, R. A. (1981) "Magic-Angle Carbon-13 Nuclear Magnetic Resonance Study of the Compatibility of Solid Polymeric Blends," *Macromolecules* **14**, 275–279.

II.F

Schaefer, J. and Stejskal, E. O. (1979) "High-Resolution C-13 NMR of Solid Polymers," *Topics in C-113 NMR Spectroscopy*, Ed. G. C. Levy, Vol. 3, Chap. 4, 283–324.

III

Spin–Spin Interactions

In this chapter, we will examine in some detail the dipole–dipole interaction, which is ubiquitous in NMR studies. We will begin with classical considerations and then develop the quantum mechanical picture.

III.A. GENERAL CONSIDERATIONS

We will begin with the general properties of the classical interaction, then justify the division into secular and non-secular parts, using the quantum mechanical form, develop some special cases, and then show how magic angle spinning (MAS) affects the dipolar-broadened NMR line width.

III.A.1. The Dipole–Dipole Interaction

Consider Fig. III.A.1: classically, the field produced by a magnetic dipole moment μ_I at a distance $\mathbf{r}$ is given by

$$\mathbf{B}_I = \nabla[\mu_I \cdot \nabla(1/r)] \tag{III.A.1}$$

where the gradient operator is defined in Appendix 1.

If we use

$$r = (x^2 + y^2 + z^2)^{1/2} \tag{III.A.2}$$

so that

$$\nabla r = [\hat{\mathbf{i}}x + \hat{\mathbf{j}}y + \hat{\mathbf{k}}z]/r = \hat{\mathbf{r}} \tag{III.A.3}$$

and

$$\nabla(1/r) = -r^{-2}\nabla r, \tag{III.A.4}$$

we can show that we obtain

$$\mathbf{B}_I = [3(\mu_I \cdot \hat{\mathbf{r}})\hat{\mathbf{r}} - \mu_I]/r^3 \tag{III.A.5}$$

for the field of dipole μ_I.

Since the energy of interaction of another dipole μ_s at a point where the magnetic field is given by $\mathbf{B}_I$ is

$$E = -\mu_S \cdot \mathbf{B}_I, \tag{III.A.6}$$

we have

$$E = -[3(\mu_I \cdot \hat{\mathbf{r}})(\mu_S \cdot \hat{\mathbf{r}}) - \mu_I \cdot \mu_S]/r^3. \tag{III.A.7}$$

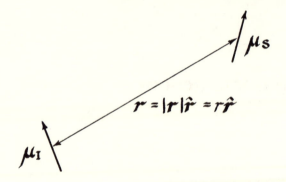

Figure III.A.1   The interaction between two magnetic dipoles.

We will use this expression for the energy of interaction between two magnetic dipoles μ_I and μ_S separated by a vector distance **r**.

The quantum mechanical analog of this expression is given by the dipolar Hamiltonian

$$H_D = -\hbar^2 \gamma_I \gamma_S r^{-3}[3(\mathbf{I}\cdot\hat{\mathbf{r}})(\mathbf{S}\cdot\mathbf{r}) - \mathbf{I}\cdot\mathbf{S}], \qquad \text{(III.A.8)}$$

where the spins are related to the magnetic moments by

$$\mu_I = \hbar\gamma_I \mathbf{I} \qquad \text{and} \qquad \mu_S = \hbar\gamma_S \mathbf{S}. \qquad \text{(III.A.9)}$$

We can write this as

$$H_D = (I_1 I_2 I_3)\begin{pmatrix} D_{11} & D_{12} & D_{13} \\ D_{21} & D_{22} & D_{23} \\ D_{31} & D_{32} & D_{33} \end{pmatrix}\begin{pmatrix} S_1 \\ S_2 \\ S_3 \end{pmatrix}, \qquad \text{(III.A.10)}$$

where

$$D = \hbar\omega_D \begin{pmatrix} 1 - 3x^2 & -3xy & -3xz \\ -3xy & 1 - 3y^2 & -3yz \\ -3xz & -3yz & 1 - 3z^2 \end{pmatrix}, \qquad \text{(III.A.11)}$$

and

$$\omega_D = \frac{\gamma_I \gamma_S \hbar}{r^3}. \qquad \text{(III.A.12)}$$

If we transform to spherical coordinates, the dipolar Hamiltonian can be written as a sum of six terms:

$$H_D = \hbar\omega_D[A + B + C + D + E + F],$$

where

$$A = (1 - 3\cos^2\theta)I_Z S_Z$$
$$B = -(\tfrac{1}{4})(1 - 3\cos^2\theta)[I_+ S_- + I_- S_+]$$
$$C = -(\tfrac{3}{2})\sin\theta\cos\theta\, e^{-i\phi}[I_Z S_+ + I_+ S_Z]$$
$$D = -(\tfrac{3}{2})\sin\theta\cos\theta\, e^{+i\phi}[I_Z S_- + I_- S_Z]$$
$$E = -(\tfrac{3}{4})\sin^2\theta\, e^{-i2\phi}[I_+ S_+]$$
$$F = -(\tfrac{3}{4})\sin^2\theta\, e^{+i2\phi}[I_- S_-]. \qquad \text{(III.A.13)}$$

(See Appendix A1 for definitions of the spin operators.)

III.A.2. The Secular and Non-Secular Parts of H_D

We assume that the two dipoles in question sit in a strong external magnetic field $\mathbf{B}_0$, taken to be in the $+z$ direction, and which is large in comparison with the local magnetic field due to the dipoles themselves. Under this condition, we can separate the dipolar Hamiltonian H_D into a secular (stationary) part and a non-secular (time-dependent) part. The first-order shift in the nth Zeeman energy level due to the perturbing dipole–dipole Hamiltonian H_D (see Section I.C.3) is simply

$$\Delta E_n = \langle n|H_D|n \rangle. \tag{III.A.14}$$

It is clear from Eq. III.A.14 that only those parts of H_D that contribute to the first-order shift are those with non-zero diagonal elements; that is, only those parts that commute with the Zeeman Hamiltonian. The Zeeman Hamiltonian for two like spins is

$$H_z = -\gamma\hbar(I_{1z} + I_{2z})B_0, \tag{III.A.15}$$

where $\mathbf{B}_0$ is taken to be in the z-direction.

Consider a term in H_D and evaluate its commutator with H_z:

$$[H_z, I_{1+}I_{2+}] = \gamma\hbar B_0(I_{2+}[I_{1z}, I_{1+}] + I_{1+}[I_{2z}, I_{2+}])$$
$$= \gamma\hbar B_0(2iI_{1+}I_{2+}) \neq 0, \tag{III.A.16}$$

so $I_{1+}I_{2+}$ does not commute with H_z. (One can reach the same conclusion by explicitly showing that $I_{1+}I_{2+}$ has no diagonal matrix elements.)

In a similar fashion, we can show that terms C, D, E, and F appearing in Eq. III.A.13 do not commute with H_z, and hence may be dropped as well. However, we find that $I_{1+}I_{2-} + I_{1-}I_{2+}$ does commute with H_z:

$$[H_z, I_{1+}I_{2-} + I_{1-}I_{2+}] = -\gamma\hbar B_0([I_{1z}, I_{1+}]I_{2-} + [I_{2z}, I_{2-}]I_{1+}$$
$$+ [I_{z-}, I_{1-}]I_{2+} + [I_{2z}, I_{2+}]I_{1-})$$
$$= -\gamma\hbar B_0(I_{1+}I_{2-} - I_{1+}I_{2-} - I_{1-}I_{2+} + I_{1-}I_{2+}) = 0. \tag{III.A.17}$$

Moreover, since

$$\mathbf{I}_1 \cdot \mathbf{I}_2 = I_{1x}I_{2x} + I_{1y}I_{2y} + I_{1z}I_{2z}$$
$$= (1/2)(I_{1+}I_{2-} + I_{1-}I_{2+}) + I_{1z}I_{2z}, \tag{III.A.18}$$

and since $I_{1z}I_{2z}$ clearly commutes with H_z, we can conclude that $\mathbf{I}_1 \cdot \mathbf{I}_2$ commutes with H_z also.

Thus for the secular part of a heteronuclear interaction ($\omega_{0I} \neq \omega_{0S}$) we have

$$H_D \approx \hbar\omega_D A = \hbar\omega_D(1 - 3\cos^2\theta)I_z S_z, \tag{III.A.19}$$

where for a homonuclear interaction ($\omega_{0I} = \omega_{0S}$) the expression becomes

$$H_D \approx \omega_D[A + D] = \omega_D(1 - 3\cos^2\theta)[I_z S_z - \tfrac{1}{4}(I_+ S_- + I_- S_+)]. \tag{III.A.20}$$

The term at the end of the right-hand side of Eq. III.A.20 represents the "flip-flop" terms, since they give rise to simultaneous opposite flips of the two nuclear spins. Note that in the homonuclear case this process conserves energy and thus is secular, whereas in the heteronuclear case it does not and is not.

We can write

$$I_+S_- + I_-S_+ = (I_x + iI_y)(S_x - iS_y) + (I_x - iI_y)(S_x + iS_y)$$

$$= 2(I_xS_x + I_yS_y) = 2(\mathbf{I}\cdot\mathbf{S} - I_zS_z), \qquad \text{(III.A.21)}$$

so that the homonuclear interaction can be written

$$H_D \approx \omega_D(1 - 3\cos^2\theta)[I_zS_z - \tfrac{1}{2}\mathbf{I}\cdot\mathbf{S} + \tfrac{1}{2}I_zS_z] \qquad \text{(III.A.22)}$$

or

$$H_D \approx \frac{\omega_D}{2}(1 - 3\cos^2\theta)[3I_zS_z - \mathbf{I}\cdot\mathbf{S}] = H_D^t, \qquad \text{(III.A.23)}$$

where the superscript "t" indicates "truncated", with non-secular terms dropped.

If we use the notation of Section I.C, we have

$$(I_+S_- + I_-S_+)|-\rangle|+) = |+\rangle|-) \qquad \text{(III.A.24)}$$

and

$$(I_+S_- + I_-S_+)|+\rangle|-) = |-\rangle|+). \qquad \text{(III.A.25)}$$

We will reintroduce the α and β notations by

$$|+\rangle|-) = |\alpha\beta\rangle, \qquad (-|\langle +| = \langle\beta\alpha|, \qquad \text{(III.A.26)}$$

and so forth, and write for the case of different types of nucleus (the heteronuclear case),

$$H_D^t = \hbar\omega_D(1 - 3\cos^2\theta)I_zS_z = KI_zS_z, \qquad \text{(III.A.27)}$$

so that

$$\langle\alpha\alpha|H_D^t|\alpha\alpha\rangle = +K/4 = \langle\beta\beta|H_D^t|\beta\beta\rangle \qquad \text{(III.A.28)}$$

and

$$\langle\alpha\beta|H_D^t|\beta\alpha\rangle = -K/4 = \langle\beta\alpha|H_D^t|\alpha\beta\rangle. \qquad \text{(III.A.29)}$$

III.A.3. The Pake Doublet

When we combine the Zeeman Hamiltonian with the dipole Hamiltonian, and look for the transitions satisfying $\Delta m_I = \pm 1$, we obtain for the NMR spectrum

$$\omega_1 = \omega_I + K/2 = \omega_I + (\omega_D/2)(1 - 3\cos^2\theta) \qquad \text{(III.A.30)}$$

and

$$\omega_2 = \omega_I - K/2 = \omega_I - (\omega_D/2)(1 - 3\cos^2\theta). \qquad \text{(III.A.31)}$$

These two lines form the "Pake doublet" for the NMR spectrum for a spin $\tfrac{1}{2}$ nucleus dipolar-coupled with a *different* type of spin $\tfrac{1}{2}$ nucleus (see Fig. III.A.2).

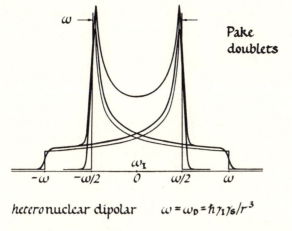

heteronuclear dipolar $\omega = \omega_D = \hbar \gamma_I \gamma_s / r^3$

homonuclear dipolar $\omega = \frac{3}{2}\omega_D$ $\omega_D = \hbar \gamma_I^2 / r^3$

quadrupolar $(I=1, \eta=0)$ $\omega = \frac{3}{4}\omega_Q$ $\omega_Q = e^2 q Q / \hbar$

Figure III.A.2   Various Pake doublets (Crystal Powder Averages).

In the homonuclear case, we should take two linear combinations of the two degenerate states so that interchanging labels of two identical spins has no observable effect; this is explicit if we write

$$|even\rangle = 2^{-1/2}(\alpha\beta + \beta\alpha) \tag{III.A.32}$$

and

$$|odd\rangle = 2^{-1/2}(\alpha\beta - \beta\alpha). \tag{III.A.33}$$

Using the homonuclear H_D^t as given in Eq. III.A.23 we find quickly that

$$\langle even|H_D^t|odd\rangle = \langle odd|H_D^t|even\rangle = 0. \tag{III.A.34}$$

We can also show easily that

$$\langle even|H_D^t|even\rangle = -K/2, \tag{III.A.35}$$

$$\langle odd|H_D^t|odd\rangle = 0, \tag{III.A.36}$$

$$\langle \alpha\alpha|H_D^t|\alpha\alpha\rangle = K/4 = \langle \beta\beta|H_D^t|\beta\beta\rangle \tag{III.A.37}$$

and that all the other matrix elements of H_D^t are zero.

The states $\alpha\alpha$, $|even\rangle$, and $\beta\beta$ can be grouped into a "triplet" state (total spin $= 1$), leaving $|odd\rangle$ as a "singlet" state (total spin $= 0$). Combining the Zeeman Hamiltonian (Eq. III.A.15) with the homonuclear H_D, we have, then, for single quantum ($\Delta m = \pm 1$) transitions

$$|even\rangle \text{ to } \alpha\alpha: \quad \omega_1 = \omega_0 + (3\omega_D/4)(1 - 3\cos^2\theta) \tag{III.A.38}$$

and

$$\beta\beta \text{ to } |even\rangle: \quad \omega_2 = \omega_0 - (3\omega_D/4)(1 - 3\cos^2\theta). \tag{III.A.39}$$

This describes the "Pake doublet" for the NMR spectrum of two identical nuclei coupled by H_D (see Fig. III.A.2). Note that the heteronuclear splitting is narrower by a factor of 2/3 from the homonuclear splitting, through the "flip-flop" terms in the latter case.

III.B. DIPOLE–DIPOLE RELATED PULSE SEQUENCES

We will now take a further step in our consideration of dipole-coupled spin systems and ask how they evolve in time, particularly in response to pulses. In general, we will use semiclassical arguments based on a vector model, and in some cases use the density matrix approach from Section I.C.6 for a more rigorous derivation of the basic results.

III.B.1. Solid Echoes

Consider the triplet state of the homonuclear two-spin case discussed in the last section. The matrix representation of the Hamiltonian is, from Eqs. III.A.12 and III.A.31–34,

$$H = -\gamma \hbar B_0 I_z + H_D^t$$

$$= \begin{pmatrix} -\hbar\omega + \hbar K/4 & 0 & 0 \\ 0 & -\hbar K/2 & 0 \\ 0 & 0 & +\hbar\omega + \hbar K/4 \end{pmatrix}. \quad \text{(III.B.1)}$$

The unitary time evolution operator (see Section I.C.6) is

$$U_H = \begin{pmatrix} \exp(i(\omega - K/4)t) & 0 & 0 \\ 0 & \exp(i(K/2)t) & 0 \\ 0 & 0 & \exp(-i(\omega + K/4)t) \end{pmatrix}. \quad \text{(III.B.2)}$$

The density matrix at $t = 0$ is (see Section I.C.6),

$$\rho(0) = \frac{1}{3} \begin{pmatrix} \exp((\hbar\omega - \hbar K/4)/kT) & 0 & 0 \\ 0 & \exp((\hbar K/2)/kT) & 0 \\ 0 & 0 & \exp((-\hbar\omega - \hbar K/4)/kT) \end{pmatrix}.$$

$$\text{(III.B.3)}$$

Since $(\hbar\omega - \hbar K/4) \ll kT$, we can expand the exponentials and approximate:

$$\rho(0) \approx \frac{1}{3} \begin{pmatrix} 1 + ((\hbar\omega - \hbar K/4)/kT) & 0 & 0 \\ 0 & 1 + ((\hbar K/2)/kT) & 0 \\ 0 & 0 & 1 - ((-\hbar\omega - \hbar K/4)/kT) \end{pmatrix}$$

$$= \frac{1}{3} + \frac{\hbar\omega}{3kT} I_z + \frac{\hbar K}{2kT} \begin{pmatrix} -1 & 0 & 0 \\ 0 & +2 & 0 \\ 0 & 0 & -1 \end{pmatrix}, \quad \text{(III.B.4)}$$

where kT is Boltzmann's constant times the absolute temperature.

We will drop the first term as time-independent, and the third term as being small relative to the second, so that

$$\rho(0) = \frac{\hbar\omega}{3kT}\begin{pmatrix} 1 & 0 & 0 \\ 0 & 0 & 0 \\ 0 & 0 & -1 \end{pmatrix} = \frac{p}{3}I_z, \tag{III.B.5}$$

where we have written $p = \hbar\omega/kT$.

The operators for $\pi/2$ pulses about $-x$ and $+y$ axes are (see Section I.C.6)

$$U-\frac{\pi}{2}x = \frac{1}{2}\begin{pmatrix} 1 & +i\sqrt{2} & -1 \\ +i\sqrt{2} & 0 & +i\sqrt{2} \\ -1 & +i\sqrt{2} & 1 \end{pmatrix} \tag{III.B.6}$$

and

$$U\frac{\pi}{2}y = \frac{1}{2}\begin{pmatrix} 1 & \sqrt{2} & 1 \\ -\sqrt{2} & 0 & \sqrt{2} \\ 1 & -\sqrt{2} & 1 \end{pmatrix}. \tag{III.B.7}$$

Refer to Fig. III.B.1; we take $\rho(0) = \rho(a)$, and use the methods of Appendix A1 to find $\rho(b)$ just after a rotation of $-\pi/2$ about the x-axis:

$$\rho(b) = \frac{p}{12}\begin{pmatrix} 1 & +i\sqrt{2} & -1 \\ +i\sqrt{2} & 0 & +i\sqrt{2} \\ -1 & +i\sqrt{2} & 1 \end{pmatrix}\begin{pmatrix} 1 & 0 & 0 \\ 0 & 0 & 0 \\ 0 & 0 & -1 \end{pmatrix}$$

$$\times \begin{pmatrix} 1 & -i\sqrt{2} & -1 \\ -i\sqrt{2} & 0 & -i\sqrt{2} \\ -1 & -i\sqrt{2} & 1 \end{pmatrix}. \tag{III.B.8}$$

Matrix multiplication leads to

$$\rho(b) = i\frac{p\sqrt{2}}{6}\begin{pmatrix} 0 & -1 & 0 \\ +1 & 0 & -1 \\ 0 & +1 & 0 \end{pmatrix} = +\frac{p}{3}I_y. \tag{III.B.9}$$

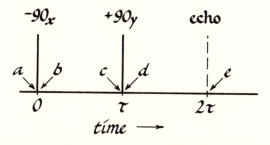

Figure III.B.1   Solid echo pulse sequence.

We now return to the time evolution operator (see Eq. III.B.39). In the frame rotating at angular frequency ω, we have

$$
U_H = \begin{pmatrix} e^{-iK\tau/4} & 0 & 0 \\ 0 & e^{iK\tau/2} & 0 \\ 0 & 0 & e^{-iK\tau/4} \end{pmatrix} = U_H^{-1}. \qquad \text{(III.B.10)}
$$

The density matrix evolves for a time t after the first pulse (see Fig. III.B.1):

$$
\rho(c) = \rho(\tau) = \frac{ip\sqrt{2}}{6} \begin{pmatrix} e^{-iK\tau/4} & 0 & 0 \\ 0 & e^{iK\tau/2} & 0 \\ 0 & 0 & e^{-iK\tau/4} \end{pmatrix} \begin{pmatrix} 0 & -1 & 0 \\ 1 & 0 & -1 \\ 0 & 1 & 0 \end{pmatrix}
$$

$$
\times \begin{pmatrix} e^{iK\tau/4} & 0 & 0 \\ 0 & e^{-iK\tau/2} & 0 \\ 0 & 0 & e^{iK\tau/4} \end{pmatrix} \qquad \text{(III.B.11)}
$$

This reduces to

$$
\rho(c) = \rho(\tau) = \frac{ip\sqrt{2}}{6} \begin{pmatrix} 0 & -e^{-i3K\tau/4} & 0 \\ e^{i3K\tau/4} & 0 & -e^{i3K\tau/4} \\ 0 & e^{-i3K\tau/4} & 0 \end{pmatrix}. \qquad \text{(III.B.12)}
$$

The transformation from $\rho(c)$ to $\rho(d)$ brought about by the second $\pi/2$ pulse is

$$
\rho(d) = \frac{ip\sqrt{2}}{24} \begin{pmatrix} 1 & \sqrt{2} & 1 \\ -\sqrt{2} & 0 & \sqrt{2} \\ 1 & -\sqrt{2} & 1 \end{pmatrix} \begin{pmatrix} 0 & -e^{-i(3/4)K\tau} & 0 \\ e^{i(3/4)K\tau} & 0 & -e^{i(3/4)K\tau} \\ 0 & e^{-i(3/4)K\tau} & 0 \end{pmatrix}
$$

$$
\times \begin{pmatrix} 1 & -\sqrt{2} & 1 \\ \sqrt{2} & 0 & -\sqrt{2} \\ 1 & \sqrt{2} & 1 \end{pmatrix}
$$

$$
= \frac{ip\sqrt{2}}{24} \begin{pmatrix} 0 & -4e^{i(3/4)K\tau} & 0 \\ 4e^{-i(3/4)K\tau} & 0 & -4e^{-i(3/4)K\tau} \\ 0 & 4e^{i(3/4)K\tau} & 0 \end{pmatrix}. \qquad \text{(III.B.13)}
$$

The density matrix now evolves for a time t to e after the p_y

pulse:

$$\rho(e) = \frac{ip\sqrt{2}}{6} \begin{pmatrix} e^{-iK\tau/4} & 0 & 0 \\ 0 & e^{+iK\tau/2} & 0 \\ 0 & 0 & e^{-iK\tau/4} \end{pmatrix} \begin{pmatrix} 0 & -e^{i(3/4)K\tau} & 0 \\ e^{i(3/4)K\tau} & 0 & -e^{-i(3/4)K\tau} \\ 0 & e^{i(3/4)K\tau} & 0 \end{pmatrix}$$

$$\times \begin{pmatrix} e^{iK\tau/4} & 0 & 0 \\ 0 & e^{-iK\tau/2} & 0 \\ 0 & 0 & e^{iK\tau/4} \end{pmatrix}$$

$$\rho(e) = \frac{ip\sqrt{2}}{6} \begin{pmatrix} 0 & -1 & 0 \\ 0 & 0 & -1 \\ 0 & 1 & 0 \end{pmatrix} = +\frac{p}{3} I_y = \rho(b), \tag{III.B.14}$$

so that we have an echo.

Note that if the second pulse had been πy, as for a Hahn spin echo, we would have had

$$U_{\pi y} = \begin{pmatrix} 0 & 0 & 1 \\ 0 & 1 & 0 \\ 1 & 0 & 0 \end{pmatrix} = U_{\pi y}^{-1}, \tag{III.B.15}$$

so that

$$\rho(d) = \frac{ip\sqrt{2}}{6} \begin{pmatrix} 0 & 0 & 1 \\ 0 & 1 & 0 \\ 1 & 0 & 0 \end{pmatrix} \begin{pmatrix} 0 & -e^{-i(3/4)K\tau} & 0 \\ e^{i(3/4)K\tau} & 0 & -e^{i(3/4)K\tau} \\ 0 & e^{-i(3/4)K\tau} & 0 \end{pmatrix} \begin{pmatrix} 0 & 0 & 1 \\ 0 & 1 & 0 \\ 1 & 0 & 0 \end{pmatrix}$$

$$= \frac{ip\sqrt{2}}{6} \begin{pmatrix} 0 & e^{-i(3/4)K\tau} & 0 \\ -e^{i(3/4)K\tau} & 0 & e^{i(3/4)K\tau} \\ 0 & -e^{-i(3/4)K\tau} & 0 \end{pmatrix} = -\rho(c). \tag{III.B.16}$$

Thus a π pulse inverts the sign of ρ, but does not alter its evolution, so it will *not* produce an echo. This is because the (non-selective) π pulse both inverts the phase errors (as in a Hahn echo) *and* inverts the local field that caused them. We have one inversion too many, so that there is no echo in this case, only for a $\pi/2$ pulse. A heteronuclear pair will not have this problem if only one spin is inverted.

III.B.2. Dipolar Order

If one takes a solid sample containing paramagnetic nuclei and puts it in a strong external magnetic field $\mathbf{B}_0$, then, in a time equal to a few spin-lattice relaxation times T_1, a nuclear magnetization will establish itself in the direction of $\mathbf{B}_0$ (see Section I.B.1); the system is said to exhibit *Zeeman order*. If, after this, magnetization is established, the sample is removed adiabatically

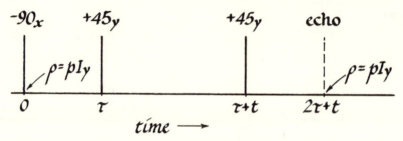

Figure III.B.2   Jeener–Broekaert sequence for producing dipolar order recovered as a solid echo.

from $\mathbf{B}_0$, the magnetization in the direction of $\mathbf{B}_0$ will no longer be present; however, if the sample is reintroduced adiabatically into $\mathbf{B}_0$, the original magnetization will reappear much more quickly than T_1.

During the period outside of $\mathbf{B}_0$, the magnetization exists in the solid sample due to alignment with the local magnetic fields, which differ from point to point in the sample. The system is said to exhibit *local order*, which can be reconverted quickly into Zeeman order. Two experimental techniques involving local order that are more elegant and easier to accomplish than physical removal and replacement of the sample with respect to $\mathbf{B}_0$ involve pulse sequences: adiabatic demagnetization in the rotating frame (ADRF) and the Jeener–Broekaert sequence.

ADRF was discussed briefly in Section II.D on cross polarization (see Fig. II.D.5). One accomplishes local order by tipping the magnetization into the x–y plane with a $\pi/2$ pulse along the x-axis and then spin-locking the magnetization by switching the phase of the rf to the y-axis; following this, one adiabatically reduces the strength of the rf amplitude B_1. The spin-locked magnetization is then oriented with respect to the local field.

The second technique involving local order is the Jeener–Broekaert sequence. One follows a $\pi/2$ pulse along the $-x$-axis with a $\pi/4$ pulse along the y-axis a time τ after the initial pulse. One then waits a further time t and administers a second $\pi/4$ pulse along the y-axis; an echo follows at time $2\tau + t$ (see Fig. III.B.2). In an analogy with the density matrix treatment of the solid echo, we can follow the density matrix of the spin system as it evolves from $r(0)$ to $r(e)$, and show that

$$\rho(e) = \rho(0), \tag{III.B.17}$$

indicating an echo.

III.C. VECTOR MODEL FOR DIPOLAR COUPLING

It is interesting that many non-trivial NMR results follow in a straight-forward fashion from a semiclassical vector model of the spins by using elementary geometry and trigonometry. We will illustrate a number of these by considering two spin-1/2 nuclei, I and S, interacting as dipoles according to Eq. III.A.8.

III.C.1. Decoupling

In the absence of an rf field $\mathbf{B}_1$ but in a strong external field $\mathbf{B}_0$, we have (see Fig. III.C.1)

$$\mathbf{I} \cdot \mathbf{S} = m_I m_S \qquad \text{(III.C.1)}$$

and

$$(\mathbf{I} \cdot \hat{\mathbf{r}})(\mathbf{S} \cdot \hat{\mathbf{r}}) = m_I m_S \cos^2 \theta, \qquad \text{(III.C.2)}$$

so that Eq. III.A.8 becomes

$$H_D^t = -(\hbar^2 \gamma_I \gamma_S m_I m_S / r^3)(3 \cos^2 \theta - 1) = -(\mu_I \mu_S / r^3)(3 \cos^2 \theta - 1). \quad \text{(III.C.3)}$$

Note that Fig. III.C.1 does not portray the precession motion of the spins (non-secular interactions).

This result is correct if $\mathbf{I}$ and $\mathbf{S}$ are unlike spins. If they are like spins, this result must be multiplied by 3/2. This additional interaction may be attributed to the presence of the "flip-flop" terms in Eq. III.A.20.

Decoupling between unlike spins $\mathbf{I}$ and $\mathbf{S}$ occurs when the $\mathbf{I}$ spin is spin-locked in the rotating x–y frame by an rf field of amplitude B_{1I}. To see this, refer to Fig. III.C.2 and note that under these conditions we have

$$\mathbf{I} \cdot \mathbf{S} = 0 \qquad \text{(III.C.4)}$$

and

$$(\mathbf{I} \cdot \hat{\mathbf{r}})(\mathbf{S} \cdot \hat{\mathbf{r}}) = m_I m_S \sin \theta \cos \theta \sin \phi. \qquad \text{(III.C.5)}$$

But

$$\phi = -\gamma_I B_0 t, \qquad \text{(III.C.6)}$$

so $\sin \phi$ has an average value of zero, and $\langle H_D^t \rangle = 0$, so $\mathbf{I}$ and $\mathbf{S}$ are decoupled.

III.C.2. Coupling in the Rotating Frame

Consider the dipolar coupling between two like spins rotating at the same rate with the same phase in the x–y plane.

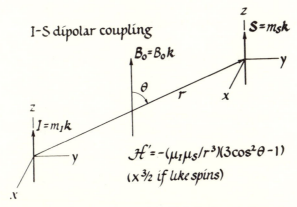

Figure III.C.1 Vector model of *I–S* dipolar coupling.

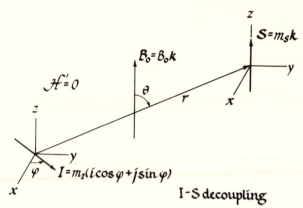

Figure III.C.2   Vector model of *I–S* dipolar decoupling.

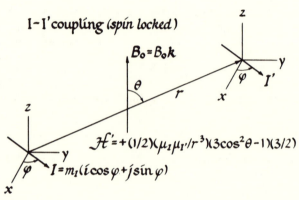

Figure III.C.3   Vector model of *I–I'* dipolar coupling in the rotating frame.

From Fig. III.C.3, we have

$$\mathbf{I} = m_I(\mathbf{i} \cos \phi + \mathbf{j} \sin \phi) \tag{III.C.7}$$

and

$$\mathbf{r} = r(\mathbf{j} \sin \theta + \mathbf{k} \cos \theta), \tag{III.C.8}$$

so that

$$\mathbf{I} \cdot \mathbf{I}' = m_I m_{I'}(\cos^2 \phi + \sin^2 \phi) = m_I m_{I'} \tag{III.C.9}$$

and

$$\mathbf{I} \cdot \mathbf{r} = m_I \sin \theta \sin \phi, \tag{III.C.10}$$

yielding

$$H_D = -(m_I m_{I'}/r^3)[3 \sin^2 \theta \sin^2 \phi - 1]. \tag{III.C.11}$$

Since $\langle \sin^2 \phi \rangle = 1/2$ and

$$\sin^2 \theta = 1 - \cos^2 \theta, \tag{III.C.12}$$

this reduces to

$$H_D^t = (1/2)(m_I m_{I'}/r^3)[3 \cos^2 \theta - 1]. \tag{III.C.13}$$

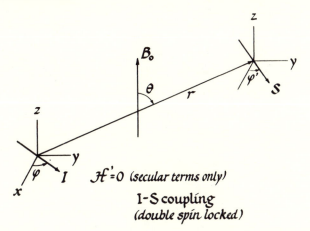

Figure III.C.4 Vector model of *I–S'* dipolar coupling in the rotating frame.

But this is just half the dipole–dipole interaction between two spins lying along the *z*-axis, as calculated above. Since these are like spins, an additional factor of 3/2 is required for numerical correctness, also as above. If *I* represents the proton spin, then this accounts for the fact that the dipolar linewidth for spin-locked protons is half the normal linewidth.

We now consider the dipolar coupling between a pair of unlike spins **I** and **S** each lying in the *x–y* plane and precessing at different rates. From Fig. III.C.4, we have

$$\mathbf{S} = m_S(\mathbf{i}\cos\phi' + \mathbf{j}\sin\phi'), \tag{III.C.14}$$

$$\mathbf{I} = m_I(\mathbf{i}\cos\phi + \mathbf{j}\sin\phi), \tag{III.C.15}$$

and

$$\mathbf{r} = (\mathbf{j}\sin\theta + \mathbf{k}\cos\theta) \tag{III.C.16}$$

$$\mathbf{I}\cdot\mathbf{S} = m_I m_S(\cos\phi\cos\phi' + \sin\phi\sin\phi'). \tag{III.C.17}$$

Moreover,

$$\mathbf{I}\cdot\mathbf{r} = m_I\sin\theta\sin\phi \tag{III.C.18}$$

and

$$\mathbf{S}\cdot\mathbf{r} = m_S\sin\theta\sin\phi', \tag{III.C.19}$$

so that

$$H_D^t = -(m_I m_S/r^3)[3\sin^2\theta\sin\phi\sin\phi' - \cos\phi\cos\phi' + \sin\phi\sin\phi']. \tag{III.C.20}$$

Since the precessional frequencies of **I** and **S** are different, each term averages to zero, so $H_D = 0$ for this case. This result seems to negate the possibility of communication (polarization transfer or cross polarization) between unlike spins, each in its own rotating frame. We will see below that this communication depends on the introduction of precessional motion about the $\mathbf{B}_1$ fields (non-secular interactions).

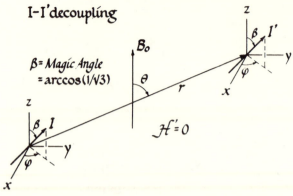

Figure III.C.5 Vector model of I–I′ dipolar decoupling in the Lee–Goldburg rotating frame.

We will now introduce a special case of the orientation of two like spins **I** and **I′** separated by a vector distance **r**. We will find that the dipole–dipole interaction vanishes for this case. How the special orientation conditions are met we will see later.

Consider two spins **I** and **I′** each making a polar angle β with the z-axis (the direction of the external magnetic field **B$_0$**) and an azimuthal angle ϕ in the x–y plane, with a vector separation **r** making an angle θ with the z-axis; we will choose the x-axis such that the azimuthal angle of **r** equals zero (see Fig. III.C.5).

$$\beta = \arccos 3^{-1/2} \tag{III.C.21}$$

is the famous magic angle encountered in magic angle spinning (MAS). As we will see in another section, spinning a solid sample about an axis making an angle β with respect to the external magnetic field leads to line narrowing by reducing the average value of dipole–dipole and chemical shift anisotropy line broadening.

From Fig. III.C.5, we have

$$\mathbf{r} = \mathbf{i} \sin \theta + \mathbf{k} \cos \theta, \tag{III.C.22}$$

$$\mathbf{I} = m_I(\mathbf{i} \cos \phi \sin \beta + \mathbf{j} \sin \phi \sin \beta + \mathbf{k} \cos \beta), \tag{III.C.23}$$

and

$$\mathbf{I'} = m_{I'}(\mathbf{i} \cos \phi \sin \beta + \mathbf{j} \sin \phi \sin \beta + \mathbf{k} \cos \beta). \tag{III.C.24}$$

Since **I** and **I′** are parallel, we have

$$\mathbf{I} \cdot \mathbf{I'} = m_I m_S. \tag{III.C.25}$$

Moreover,

$$(\mathbf{I} \cdot \mathbf{r})/m_I = (\mathbf{I'} \cdot \mathbf{r})/m_S = \sin \theta \cos \phi \sin \beta + \cos \theta \cos \beta \tag{III.C.26}$$

so that

$$(\mathbf{I} \cdot \mathbf{r})(\mathbf{I'} \cdot \mathbf{r})/m_I m_S = \sin^2 \theta \cos^2 \phi \sin^2 \beta + \cos^2 \theta \cos^2 \beta$$
$$+ 2 \sin \theta \cos \theta \cos \phi \sin \beta \cos \beta. \tag{III.C.27}$$

Now, $\cos \beta = 3^{-1/2}$ and $\sin \beta = (2/3)^{1/2}$, so

$$(\mathbf{I} \cdot \mathbf{r})(\mathbf{I}' \cdot \mathbf{r})/m_I m_S = (2/3) \sin^2 \theta \cos^2 \phi + (1/3) \cos^2 \theta$$
$$+ (3^{1/2}/3) \sin \theta \cos \theta \cos \phi. \qquad (\text{III.C.28})$$

As ϕ varies rapidly and uniformly as $\mathbf{I}$ and $\mathbf{I}'$ precess about the direction of $\mathbf{B}_0$, we have

$$\langle \cos^2 \phi \rangle = 1/2 \qquad \text{and} \qquad \langle \cos \phi \rangle = 0, \qquad (\text{III.C.29})$$

so

$$(\mathbf{I} \cdot \mathbf{r})(\mathbf{I}' \cdot \mathbf{r})/m_I m_S = (1/3)(\sin^2 \theta + \cos^2 \theta) = 1/3. \qquad (\text{III.C.30})$$

Combining the previous equations, we obtain $H_D^t = 0$ for this case.

One way to achieve the condition for homonuclear decoupling as just described is a modification of the spin-locking procedure described earlier due to Lee and Goldburg. In off-resonance spin-locking, one uses as the frequency of the locking signal not γB_0, but a frequency $\omega = \omega_0 + \delta\omega$ where the frequency offset $\delta\omega$ is defined by an equation, to be derived shortly, that locks the spins at the magic angle with respect to the z-axis. We require that the off-resonance effective field along the z-axis satisfies

$$\tan \beta = B_1/B_{\text{off}} = (2)^{1/2} \qquad (\text{III.C.31})$$

where

$$\mathbf{B}_{\text{off}} = \mathbf{B}_0 + \omega/\gamma. \qquad (\text{III.C.32})$$

(See Fig. III.C.6 and note that if $\tan \beta = (2)^{1/2}$, then $\cos \beta = (3)^{-1/2}$, so that β is the magic angle.)

Since $\omega_0 = -\gamma B_0$, we have

$$\delta\omega = \gamma B_1 (2)^{-1/2}. \qquad (\text{III.C.33})$$

Another technique to achieve homonuclear decoupling is to use a sequence of $\pi/2$ pulses (called WAHUHA after Waugh, Huber and Haeberlen) that have the effect of putting the nuclear magnetization for equal amounts of time along the x, y, and z axes in the rotating frame. The sequence of pulses is given by

$$[t - (\pi/2)_x - t - (\pi/2)_{-y} - 2t - (\pi/2)_y - t - (\pi/2)_{-x} - t]_n, \qquad (\text{III.C.34})$$

where the subscript n indicates that the pulse should be repeated (see Fig. III.C.7). Thus, in a crude sense, the time average orientation of the

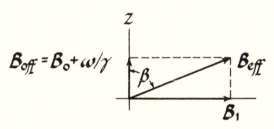

Figure III.C.6 Use of frequency offset to spin lock out of the x–y plane.

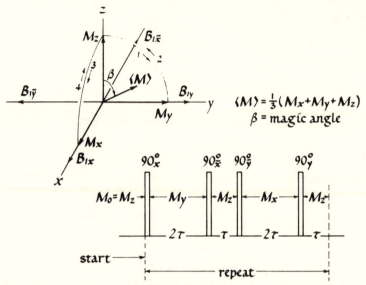

$\langle M \rangle = \frac{1}{3}(M_x + M_y + M_z)$

β = magic angle

Figure III.C.7 Vector model of WaHuHa sequence (somewhat spurious).

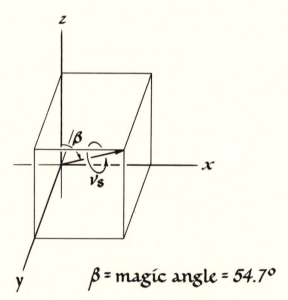

β = magic angle = 54.7°

Figure III.C.8 Magic angle geometry.

magnetization is at the magic angle with respect to the z-axis (see Fig. III.C.8), suggesting that this sequence of pulses might lead to homonuclear decoupling. This is in fact the case, but the rigorous proof is much more involved.

Finally, we consider what the I–S dipolar coupling looks like under the conditions, just discussed, under which the I–I' dipolar coupling vanishes. We have the S-spin along the z-axis with the I-spin precessing at the

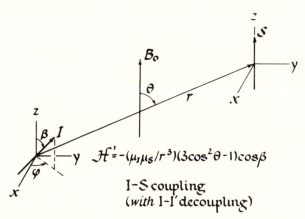

$$\mathcal{H}' = -(\mu_I \mu_S/r^3)(3\cos^2\theta - 1)\cos\beta$$

I–S coupling
(with I–I' decoupling)

Figure III.C.9 Vector model of *I–S* dipolar coupling in the rotating frame, in the presence of *I–I'* decoupling.

resonance frequency at the magic angle β with respect to the z-axis (see Fig. III.C.9). Again, we choose the y-axis so that the internuclear vector **r** is given by

$$\mathbf{r} = r(\mathbf{j}\sin\theta + \mathbf{k}\cos\theta). \tag{III.C.35}$$

Furthermore,

$$\mathbf{S} = m_S\mathbf{k} \tag{III.C.36}$$

and

$$\mathbf{I} = m_I(\mathbf{i}\cos\phi\sin\beta + \mathbf{j}\sin\phi\sin\beta + \mathbf{k}\cos\beta) \tag{III.C.37}$$

so that

$$\mathbf{I}\cdot\mathbf{S} = m_I m_S \cos\beta, \tag{III.C.38}$$

$$\mathbf{S}\cdot\mathbf{r} = m_S \cos\theta, \tag{III.C.39}$$

and

$$\mathbf{I}\cdot\mathbf{r} = m_I(\sin\theta\sin\phi\sin\beta + \cos\theta\cos\beta). \tag{III.C.40}$$

As **I** precesses rapidly, $\langle\sin\phi\rangle = 0$, so

$$3(\mathbf{I}\cdot\mathbf{r})(\mathbf{S}\cdot\mathbf{r}) - \mathbf{I}\cdot\mathbf{S} = m_I m_S \cos\beta(3\cos^2\theta - 1). \tag{III.C.41}$$

We have, then,

$$H_D = -(m_I m_S/r^3)(3\cos^2\theta - 1)\cos\beta. \tag{III.C.42}$$

But, from Eq. III.C.31, $\cos\beta = 3^{-1/2}$, so

$$H_D = -3^{-1/2}(m_S m_I/r^3)(3\cos^2\theta - 1). \tag{III.C.43}$$

That is, the heteronuclear dipolar coupling is scaled down by a factor $3^{-1/2}$.

We now return to the problem of cross polarization.

Consider Fig. III.C.10, which represents a spin-locked (see Section II.D) spin, precessing about the x-axis in the rotating frame at an angle χ. Using the old vector model, we have, for the magnitude of the spin vector I,

$$\mathbf{I}\cdot\mathbf{I} = \sqrt{I(I+1)} \tag{III.C.44}$$

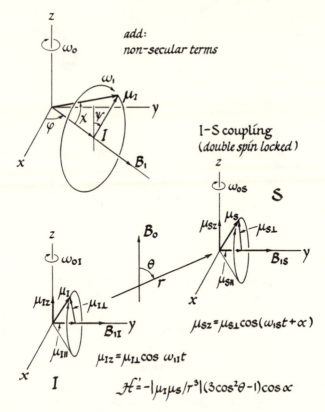

Figure III.C.10 Vector model of *I–S* dipolar coupling in the rotating frame, with precessional terms added.

and
$$m_I = \pm I = \pm \tfrac{1}{2} \tag{III.C.45}$$
so that
$$\cos \chi = \pm\sqrt{3}/3. \tag{III.C.46}$$
In the rotating frame, for $\phi = 0$, we can see that
$$\mathbf{I} = m_I[\mathbf{i} \cos \chi + \mathbf{j} \sin \chi \sin \psi + \mathbf{k} \sin \chi \cos \psi], \tag{III.C.47}$$
where
$$\psi = \omega_{1I}t. \tag{III.C.48}$$
If the overall precession is restored, so that
$$\phi = \omega_{0I}t, \tag{III.C.49}$$
this becomes
$$
\begin{aligned}
\mathbf{I} = m_I[&\mathbf{i} \cos \chi \cos \phi + \mathbf{j} \cos \chi \sin \phi \\
&+ \mathbf{i} \sin \chi \sin \psi(-\sin \phi) + \mathbf{j} \sin \chi \sin \psi \cos \phi + \mathbf{k} \sin \chi \cos \psi] \\
= m_I[&\mathbf{i}(\cos \chi \cos \phi - \sin \chi \sin \psi \sin \phi) \\
&+ \mathbf{j}(\cos \chi \sin \phi + \sin \chi \sin \psi \cos \phi) + \mathbf{k} \sin \chi \cos \psi].
\end{aligned}
\tag{III.C.50}
$$

For a second, unlike spin S, we have a similar equation but with

$$\phi' = \omega_{0S}t \quad \text{and} \quad \psi' = \omega_{1S}t + \alpha. \quad \text{(III.C.51)}$$

Using the Hartmann–Hahn condition, the fact that the time-average of the cosine function is zero, and a good deal of tedious algebra, we find that the coupling Hamiltonian becomes

$$H_D = -(\hbar^2\gamma_I\gamma_S/r^3)[\tfrac{1}{2}\cos\alpha(m_I\sin\chi)(m_S\sin\chi')](3\cos^3\theta - 1). \quad \text{(III.C.52)}$$

The spins couple, then, providing the phase angle α satisfies

$$\cos\alpha \neq 0. \quad \text{(III.C.53)}$$

This reduces to

$$H_D = -(\hbar^2\gamma_I\gamma_S/r^3)\left(\frac{\cos\alpha}{4}\right)(3\cos^2\theta - 1)$$

$$= -|\mu_I\mu_S/r^3|(\cos\alpha)(3\cos^2\theta - 1). \quad \text{(III.C.54)}$$

For $\alpha = 0$, this is identical with simple dipolar coupling. Why?

We have $m_\perp > m_I$, so that even though the coupling is only part-time, it is stronger when it does act.

This picture of the coupling that enables CP transfer can also be used to understand $T_{1\rho}(S)$, $T_{1\rho}(I)$, and spin diffusion.

When both the I and S spins are spin locked so that their z-components have a common time dependence, they can communicate even though they are unlike spins. Consider $T_{1\rho}(S)$: when the S spins are spin locked and the I spins are aligned along the z-axis there is no common time dependence of the z-components, that is, unless something causes the I-spin z-component to fluctuate with spectral density at ω_{1S}. This fluctuation can arise from motion of the I and S spins relative to one another (spin-lattice mechanism) or from the I spin undergoing mutual spin flips with other coupled I spins (spin–spin mechanism). The interaction that enables coupling can be written $b(r, \theta)I_zS_\pm$. Spin-lattice relaxation arises from the time dependence of $b(r, \theta)$ and spin–spin relaxation from the time dependence of I_z. Generally, the spectral content of $I_z(t)$ drops rapidly over $\sim 30\,\text{kHz}$ so that in a system with adequate motion, spin-lattice effects begin to dominate when $\omega_{1S} > 30\,\text{kHz}$. This picture should be viewed as a combination of Fig. III.C.2 (with I and S reversed and Fig. III.C.10 non-secular variation included).

There is no problem for two like spins to communicate when they are spin locked since their z-components naturally have a common time dependence. (The Hartmann–Hahn condition is always met.) This rapid communication is not usually observed as a polarization transfer (at least not in homogeneous systems) but instead gives rise to spin diffusion and the I-spin fluctuation spectrum. However, since no lattice energy is involved, these mutual spin flips are of the up-down to down-up variety. How then does $T_{1\rho}(I)$, a spin lattice relaxation, occur? Relative motion of the spins (at frequency ω_m) modulates the "perfect" $\omega_{1I} = \omega_{1I}'$ match to the "less-than-perfect" ω_{1I} versus $\omega_{1I}' \pm \omega_m$. At first, motion reduces the rate of communication between coupled spins (and the I-spin line width). But when $\omega_m = 2\omega_{1I}$, then we have ω_{1I} versus $\omega'_{1I} + 2\omega_{1I} = 3\omega_{1I}$ (ineffective) and ω_{1I} versus $\omega'_{1I} - 2\omega_{1I} = -\omega_{1I}$ which is effective since the sign of the frequency is unimportant. This is why $T_{1\rho}(I)$ is sensitive to motions near $2\omega_{1I}$ while $T_{1\rho}(S)$ detects motions (spin–spin or spin–lattice) near ω_{1S}.

$T_{1\rho}(I)$ and $T_{1\rho}(S)$ are useful in the characterization of molecular motion. The CP/MAS experiment requires the manipulation of both the abundant spin I (usually ^{1}H, but sometimes ^{19}F) and the rare spin S (often ^{13}C) along the radiofrequency fields appropriate to each. These manipulations can involve several rotating-frame relaxation times, most often $T_{1\rho}(^1$H) and $T_{1\rho}(^{13}$C), that are sensitive to the frequency and amplitude of the motion of the spin system being studied. Frequencies in the range from a few kiloHertz to a few hundred kiloHertz are the most important in determining these relaxation times; near room temperature, this frequency range reflects, for instance, the motion of short segments of polymer main chains, as well as side-chain motion. Anything that alters these motions may also alter one of the rotating-frame relaxation times.

$T_{1\rho}(^1$H) is different from $T_{1\rho}(^{13}$C) in that $T_{1\rho}(^1$H) is sensitive to the motion of the ^{1}H system averaged by spin diffusion over a short distance while $T_{1\rho}(^{13}$C) relaxation reflects the motion of individual ^{13}C spins. Thus the ^{1}H relaxation is sensitive to macroscopic variations in motion while the ^{13}C relaxation is site specific. Figure III.C.11 depicts a group of spins relaxing

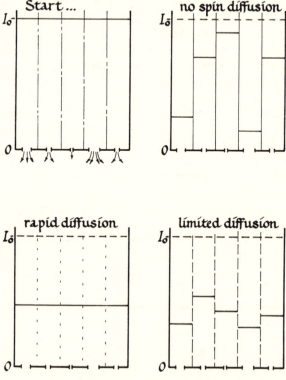

Figure III.C.11 Schematic representation of the effect of spin diffusion in heterogeneous systems with varying amounts of communication between regions.

at different rates as liquid reservoirs draining at different rates. The effect of different rates of spin diffusion (different rates of leakage between reservoirs) can be seen as the reservoirs drain. Together, these two relaxation processes not only detect changes in molecular motion but also motional inhomogeneity. Even in fundamentally homogeneous systems, these two relaxation times still behave differently in that $T_{1\rho}(^1\mathrm{H})$ tends to be dominated by the most efficient site of relaxation even if it is a minority site. On the other hand, $T_{1\rho}(^{13}\mathrm{C})$ reports the most common relaxation rate and de-emphasizes minor sources of relaxation. Since each $T_{1\rho}(^{13}\mathrm{C})$ measured is associated with a particular carbon, it is totally insensitive to motion of any other chemically distinct species even if that species may dominate $T_{1\rho}(^1\mathrm{H})$.

When relaxation is altered by changes in a system it may be due to either or both of two fundamentally different changes: either the rate of motion may change in relation to the characteristic frequency of the measurement, or the character of the motion (amplitude or type) may change. These different sources of change can be sorted out by studying relaxation as a function of experimental parameters or by the introduction of still more kinds of relaxation phenomena. There are also a number of lineshape measurements which are affected by motion.

BIBLIOGRAPHY

Articles

III.A.3

Andrew, E. R. and Bersohn, R. (1950) "Nuclear Magnetic Resonance Line Shape for a Triangular Configuration of Nuclei," *J. Chem. Phys.* **18**, 159–161.

Pake, G. E. (1948) "Nuclear Resonance Absorption in Hydrated Crystals: Fine Structure of the Proton Line," *J. Chem. Phys.* **16**, 327–336.

Van Vleck, J. H. (1948) "The Dipolar Broadening of Magnetic Resonance Lines in Crystals," *Phys. Rev.* **74**, 1168–1183.

III.B.1

Pines, A., Rhim, W. K., and Waugh, J. S. (1972) "Homogeneous and Inhomogeneous Nuclear Spin Echoes in Solids," *J. Magn. Reson.* **6**, 457–465.

Powles, J. G. and Mansfield, P. (1962) "Double-Pulse Nuclear-Resonance Transients in Solids," *Phys. Lett.* **2**, 58–59.

Powles, J. G. and Strange, J. H. (1963) "Zero Time Resolution Nuclear Magnetic Resonance Transients in Solids," *Proc. Phys. Soc., London* **82**, 6–15.

III.B.2

Slichter, C. P. and Holton, W. C. (1961) "Adiabatic Demagnetization in a Rotating Reference System," *Phys. Rev.* **122**, 1701–1708.

III.C.2

Cheung, T. T. P. and Yaris, R. (1980) "Dynamics of Cross Relaxation in Modulated Systems: Application to Nuclear Magnetic Double Resonance of Glassy Polymers," *J. Chem. Phys.* **72**, 3604–3616.

Goldburg, W. I. and Lee, M. (1963) "Nuclear Magnetic Resonance Line Narrowing By a Rotating rf Field," *Phys. Rev. Lett.* **11**, 255–260.

Haeberlen, U. and Waugh, J. S. (1968) "Coherent Averaging Effects in Magnetic Resonance," *Phys. Rev.* **175**, 453–467.

Jones, G. P. (1966) "Spin-Lattice Relaxation in the Rotating Frame: Weak-Collision Case," *Phys. Rev.* **148**, 332–335.

Lee, M. and Goldburg, W. I. (1965) "Nuclear-Magnetic-Resonance Line Narrowing by a Rotating rf Field," *Phys. Rev.* **140**, A1261–A1271.

Schaefer, J., Sefcik, M. D., Stejskal, E. O., and McKay, R. A. (1984) "Carbon-13 $T_{1\rho}$ Experiments on Solid Polymers Having Tightly Spin-Coupled Protons," *Macromolecules* **17**, 1118–1124.

Schaefer, J., Stejskal, E. O., and Buchdahl, R. (1977) "Magic-Angle ^{13}C NMR Analysis of Motion in Solid Glassy Polymers" *Macromolecules* **10**, 384–405.

Slichter, C. P. and Alion, D. (1964) "Low-Field Relaxation and the Study of Ultraslow Atomic Motions by Magnetic Resonance," *Phys. Rev.* **135**, A1099–A1110.

Stejskal, E. O., Schaefer, J., and Steger, T. R. (1979) "High-Resolution C-13 NMR in Solids," *Faraday Symp. Chem. Soc.* **13**, 56–62.

Stejskal, E. O., Schaefer, J., Sefcik, M. D., and McKay, R. A. (1981) "Magic-Angle Carbon-13 Nuclear Magnetic Resonance Study of the Compatibility of Solid Polymeric Blends," *Macromolecules* **14**, 275–279.

VanderHart, D. L. and Garroway, A. N. (1979) "^{13}C NMR Rotating Frame Relaxation in a Solid with Strongly Coupled Protons: Polyethylene," *J. Chem. Phys.* **71**, 2773–2787.

Waugh, J. S. and Huber, L. M. (1967) "Method for Observing Chemical Shifts in Solids," *J. Chem. Phys.* **47**, 1862–1863.

Waugh, J. S., Huber, L. M., and Haeberlen, U. (1968) "Approach to High-Resolution NMR in Solids," *Phys. Rev. Lett.* **20**, 180–182.

Waugh, J. S., Wang, C. H., Huber, L. M., and Vold, R. L. (1968) "Multiple-Pulse NMR Experiments," *J. Chem. Phys.* **48**, 662–670.

IV

Magic Angle Sample Spinning

In this chapter, we treat in somewhat more experimental and theoretical depth the MAS technique (see Section II.C).

IV.A. INTRODUCTION

As discussed in Section II.A, the NMR absorption line in solid samples in the typical experiment has a linewidth of several tens of kiloHertz, owing largely to the variation in local magnetic field due to the dipole–dipole interaction between nuclei, but also due to the anisotropy of the chemical shift. In liquids, rapid and random molecular rotation has the effect of averaging the dipole–dipole interaction and chemical shift anisotropy to zero, leading to sharp lines (several orders of magnitude narrower than for the case of solids). This makes possible in NMR experiments on liquid samples the observation of hyperfine interactions such as isotropic chemical shifts, electron-coupled spin–spin splitting, and so forth.

Mechanical spinning of a *solid* sample about an axis making a certain angle with the direction of the external magnetic field can lead to a dramatic reduction of the NMR linewidth due to the dipole–dipole interaction, suggesting an approach to the observation of spectral structure comparable to that seen with liquid samples. In practice, the effect of spinning on the dipolar sources of broadening is disappointing because of the homogeneous character imparted to that broadening by the I–I fluctuations of the abundant spins. Heteronuclear (and homonuclear) decoupling is more effective. It is, however, the case that such spinning also reduces broadening due to anisotropy in chemical shift tensors and the anisotropic part of the electron-coupled spin–spin interactions. It is for these interactions that spinning is most useful.

IV.B. THE HAMILTONIANS OF THE NUCLEAR SPIN INTERACTIONS

In this section, we examine the effect of solid sample spinning on the Hamiltonians through which nuclear magnetic moments interact.

IV.B.1. The Dipolar Hamiltonian

The physical origins of the line broadening due to the dipole–dipole interaction is a local field at the site of the nucleus due to each of the neighboring nuclei, which will cause the Zeeman energy levels to shift slightly, corresponding to a slight shift in NMR frequency. This local field will vary from point to point around the sample, as more or fewer neighboring nuclei have spin up or spin down. This effect can be treated in a formal theoretical fashion by using time-independent perturbation theory, in which the dipole–dipole Hamiltonian shifts the Zeeman levels of the nuclear spin in the external magnetic field.

From Eq. III.A.21 and Fig. II.C.1, we have, for the secular part of the dipolar Hamiltonian for two nuclei, the expression

$$H_D^t = \sum_{i<j} (\hbar^2/2)\gamma_i\gamma_j r_{ij}^{-3}(3\cos^2\theta_{ij} - 1)(\mathbf{I}_i\cdot\mathbf{I}_j - 3I_{iz}I_{jz}), \quad \text{(IV.B.1)}$$

In an isotropic liquid, $\cos^2\theta_{ij}$ takes on its random time average value of $1/3$ (see Appendix A2), so H_D^t averages to zero due to rapid and random molecular rotation, with the result that dipolar broadening of the NMR line vanishes.

In the case of a rigid solid, however, we must be careful. If we spin the solid about an axis making an angle β with respect to the external magnetic field at an angular frequency ω_r, we can show with reference to Fig. II.C.1 and some trigonometry, that

$$\cos\theta_{ij}(t) = \cos\beta\cos\beta'_{ij} + \sin\beta\sin\beta'_{ij}\cos(\omega_r t + \phi'_{ij}), \quad \text{(IV.B.2)}$$

where β'_{ij} is the angle between $\mathbf{r}_{ij}$ and the axis of rotation and ϕ_{ij} is the initial azimuth angle of $\mathbf{r}_{ij}$. Squaring Eq. IV.B.2 and taking an average over time (see Appendix A2) leads to

$$\overline{\cos^2\theta_{ij}} = (1/6)(3\cos^2\beta - 1)(3\cos^2\beta'_{ij} - 1) + 1/3. \quad \text{(IV.B.3)}$$

Combining this with Eq. IV.B.1 leads to the following expression for the time-averaged value of the truncated dipolar Hamiltonian:

$$\overline{H_D^t} = (\hbar^2/4)(3\cos^2\beta - 1)\sum_{i<j}\gamma_i\gamma_j r_{ij}^{-3}(3\cos^2\beta'_{ij} - 1)(\mathbf{I}_i\cdot\mathbf{I}_j - 3I_{iz}I_{jz}).$$

$$\text{(IV.B.4)}$$

It is clear that by choosing β so that $3\cos^2\beta - 1 = 0$, we have $\overline{H_D^t} = 0$, so that dipolar broadening vanishes. That is, if one rotates a solid sample about an axis making an angle

$$\beta_M = \arccos(1/3)^{1/2} = 54.7 \text{ degrees} \quad \text{(IV.B.5)}$$

with the direction of the external magnetic field $\mathbf{B}$, then the dipolar broadening of the NMR line disappears and one obtains the same resolution as for liquid samples (and for somewhat the same reason). So startling is this result, and so profound the effect of sample spinning, that β_M is called the "magic angle", and the experimental technique "magic angle spinning"

(MAS) or also "magic angle sample spinning" (MASS) or, less frequently, "magic angle rotation" (MAR).

IV.B.2. Chemical Shift Anisotropy

Another source of NMR line broadening in solid samples is chemical shift anisotropy (CSA). In Section I.A.5 we described how an absorption line can be shifted by a local magnetic field caused by electronic currents induced by turning on the strong external magnetic field. For molecules executing rapid and random rotation in liquids, this local magnetic field B', proportional to the external field B, is written

$$B' = -\sigma B, \tag{IV.B.6}$$

where σ is the chemical shielding parameter.

In solids, it may be necessary to write this

$$\mathbf{B'} = -\boldsymbol{\sigma} \cdot \mathbf{B}, \tag{IV.B.7}$$

where $\boldsymbol{\sigma}$ is now a second rank tensor. That is, the local chemical shift field may differ according to the direction of $\mathbf{B}$. Consider, for example, a planar molecule containing one or more aromatic rings. If the external magnetic field $\mathbf{B}$ is perpendicular to the ring, the electronic current induced around the ring by turning $\mathbf{B}$ on will produce a different local field from that induced if $\mathbf{B}$ is parallel to the plane of the ring.

If we take $\mathbf{B}$ to be in the z-direction, and neglect terms including I_x and I_y since they do not commute with the Zeeman Hamiltonian (see Section III.A), we obtain

$$H_C = \sum_i \hbar \sigma_{izz} I_{iz} B \tag{IV.B.8}$$

for the chemical shift Hamiltonian; the subscript i refers to a particular nucleus. If the symmetric part of the tensor $\boldsymbol{\sigma}_i$ has principal values σ_{i1}, σ_{i2} and σ_{i3} along its principal axes, and if θ_{i1}, θ_{i2} and θ_{i3} are the three angles made between each of these axes and the direction of $\mathbf{B}$, then

$$\sigma_{izz} = \sigma_{i1} \cos^2 \theta_{i1} + \sigma_{i2} \cos^2 \theta_{i2} + \sigma_{i3} \cos^2 \theta_{i3}. \tag{IV.B.9}$$

Now for rapid and random rotation, each angle in Eq. IV.B.9 varies randomly, so the time average of σ_{izz} is given by

$$\sigma_{izz} = (1/3)(\sigma_{i1} + \sigma_{i2} + \sigma_{i3}) = (1/3) \operatorname{Tr} \boldsymbol{\sigma}_i = \sigma_i, \tag{IV.B.10}$$

since $\overline{\cos^2 \theta} = 1/3$ (see Appendix A2), and we have an isotropic chemical shielding parameter.

In a solid sample, however, differing nuclear sites will have, in general, different values for the angles appearing in Eq. IV.B.9, so there will be a distribution of local fields and, hence, a broadened line. If, however, the solid sample is rotated at an angular frequency ω_r about an axis making an angle β with respect to the magnetic field $\mathbf{B}_0$ and angles χ_{i1}, χ_{i2}, and χ_{i3} with respect to the principal axes of the sample we have

$$\cos \theta_{ip} = \cos \beta \cos \chi_{ip} + \sin \beta \sin \chi_{ip} \cos(\omega_r t + \psi_{ip}) \quad \text{(IV.B.11)}$$

for the pth principal axis; ψ_{ip} is the initial azimuthal angle.

Using (Eqs. IV.B.9 and IV.B.11, we obtain

$$\overline{\sigma_{izz}} = (1/2)(\sin^2 \beta)(\sigma_{i1} + \sigma_{i2} + \sigma_{i3}) + (1/2)(3 \cos^2 \beta - 1) \sum_p \sigma_{ip} \cos^2 \chi_{ip}.$$
$$\text{(IV.B.12)}$$

If $\beta = \beta_M$, the same magic angle as for the dipolar Hamiltonian, then Eq. IV.B.12 reduces to Eq. IV.B.10, and the MAS solid case reduces to the liquid case.

IV.B.3. Spin–Spin Coupling

An electron with non-zero density at the sites of two nuclei can interact with both through the Fermi contact hyperfine interaction, and thus provide an intermediary way through which the nuclei can interact with one another; this interaction is usually called spin–spin coupling, or J coupling. The form of the interaction is written

$$H_J = h \sum_{i<j} \mathbf{I}_i \cdot \mathbf{J}_{ij} \cdot \mathbf{I}_j, \quad \text{(IV.B.13)}$$

where the sum is over all pairs of nuclei i and j; the interaction tensor $\mathbf{J}_{ij}$ is traditionally expressed in Hertz. We will write the second rank tensor $\mathbf{J}$ as a sum of three parts: a scalar times the unit tensor, a symmetric part, and an antisymmetric part. If we write the transpose of $\mathbf{J}$ as $\tilde{\mathbf{J}}$, then clearly

$$\mathbf{J}_{ij}^A = (\mathbf{J}_{ij} - \tilde{\mathbf{J}}_{ij})/2 \quad \text{(IV.B.14)}$$

is antisymmetric (its transpose is its negative), and

$$\mathbf{J}_{ij}^s = (\mathbf{J}_{ij} + \tilde{\mathbf{J}}_{ij})/2 - J_{ij}\mathbf{1} \quad \text{(IV.B.15)}$$

is symmetric (its transpose equals itself). Moreover, we can combine Eqs. IV.B.14 and IV.B.15 to obtain

$$\mathbf{J}_{ij} = J_{ij}\mathbf{1} + \mathbf{J}_{ij}^s + \mathbf{J}_{ij}^A. \quad \text{(IV.B.16)}$$

We will take

$$J_{ij} = (\text{Trace } \mathbf{J}_{ij})/3. \quad \text{(IV.B.17)}$$

$\mathbf{J}_{ij}^A$ is traceless, since each diagonal element must be zero, and, from Eqs. IV.B.14 and IV.B.17, Trace $\mathbf{J}_{ij}^s = 0$ also. We have, then, for the isotropic average of $\mathbf{J}_{ij}$,

$$\overline{\mathbf{J}_{ij}} = (\text{Trace } \mathbf{J}_{ij})/3 = J_{ij}(\text{Trace } \mathbf{1})/3 = J_{ij}. \quad \text{(IV.B.18)}$$

Hence

$$\overline{H_j} = h \sum_{i<j} J_{ij}\mathbf{I}_i \cdot \mathbf{I}_j, \quad \text{(IV.B.19)}$$

which is simply the result from NMR in the liquid state.

Now, consider specifically the effect of rotation of a solid sample on the three terms in Eq. IV.B.16. To begin with, there is, of course, no effect on the scalar term $J_{ij}\mathbf{1}$. The symmetric tensor for the dipolar interaction is axially symmetric, since it does not depend on the azimuthal angle ϕ, but this is not necessarily the case for $\mathbf{J}^s_{ij}$, the symmetric part of the $\mathbf{J}$ tensor. Consider the contribution of one pair of nuclei to H_J from the symmetric part of the interaction:

$$\mathbf{I}_1 \cdot \mathbf{J}^s \cdot \mathbf{I}_2 = \sum_i \sum_j J_{ij} I_{1i} I_{2j} \qquad \text{(IV.B.20)}$$

where i and j run from 1 to 3 and refer to the Cartesian components x, y, and z of the spin vectors.

Moreover, we can write

$$\mathbf{I}_1 \cdot \mathbf{J}^s \cdot \mathbf{I}_2 = \sum_i \sum_j \sum_p \cos \theta_{ij} \cos \theta_{jp} J_p I_{1i} I_{2j} \qquad \text{(IV.B.21)}$$

where θ_{ip} is the angle between the i-axis and the pth principal axis of $\mathbf{J}^s$, where J_p is the pth principal value of $\mathbf{J}^s$. The triple sum on the right-hand side has twenty-seven terms. It can be shown that if terms are dropped for the same reasons given in justifying the truncation of H_D (dropping those terms that do not commute with H_z), we obtain

$$\mathbf{I}_1 \cdot \mathbf{J}^s \cdot \mathbf{I}_2 = -J_3[(3 \cos^2 \theta_{z3} - 1) + \eta_J(\cos^2 \theta_{z2} - \cos^2 \theta_{z1})]$$
$$\times [\mathbf{I}_1 \cdot \mathbf{I}_2 - 3I_{1z}I_{2z}]/4 \qquad \text{(IV.B.22)}$$

where we have taken $J_3 > J_2 > J_1$, and

$$\eta_J = (J_2 - J_1)/J_3. \qquad \text{(IV.B.23)}$$

Using the same methods as we did earlier in finding the time averages of the direction cosines when the sample undergoes spinning, we can show that

$$\overline{\cos^2 \theta_{zp}} = (3 \cos^2 \beta - 1)(3 \cos^2 \mu_p - 1)/6 + 1/3, \qquad \text{(IV.B.24)}$$

where β is the angle between the rotor axis and the z-axis, and the μ_p are the angles between the principal axes of $\mathbf{J}^s$ and the spinning axis. If $\beta = \beta_M$, Eq. IV.B.24 shows that Eq. IV.B.22 averages to zero upon MAS. However, in general, the antisymmetric part of H_J does not average to zero.

IV.B.4. The Electric Quadrupole Interaction

It can be shown in a fashion similar to the treatments just given that if the nucleus in question has a spin greater than 1/2, and hence has an electric quadrupole moment, the interaction H_Q of that moment with the electric field gradient at the site of that nucleus has a time average of zero under MAS, as long as first-order perturbation theory is adequate; to narrow the second-order effects, a more elaborate experimental technique is required (see Section IV.G).

IV.C. SIDEBAND INTENSITIES

In the last section we considered the time average part of the perturbing Hamiltonians, and showed how MAS can greatly reduce the linewidths for solid samples due to those interactions. We now explore how the system develops as a function of time.

IV.C.1. Spinning Sidebands Arising from CSA Through MAS

We begin by writing the CSA interaction (see Eq. IV.B.8) as a function of time arising because of sample rotation:

$$H_c = \sum_i \sum_{p=1}^{3} I_{iz} B \sigma_{ip} [\cos^2 \beta \cos^2 \chi_{ip}$$

$$+ 2 \cos \beta \sin \beta \cos \chi_{ip} \sin \chi_{ip} \cos(\omega_r t + \psi_{ip})$$

$$+ \sin^2 \beta \sin^2 \chi_{ip} \cos^2(\omega_r t + \psi_{ip})], \qquad \text{(IV.C.1)}$$

where we used Eqs. IV.B.9 and IV.B.11.

Notice that the interaction contains terms that are periodic with frequency $f_r = \omega_r/2\pi$ (and also $2f_r$). It can be shown (see Appendix A3) that this time-dependence gives rise in the FID to a series of echoes separated by a time interval $2\pi/\omega_r = f_r^{-1}$. These echoes are related, in some sense, to the standard spin echoes due to forced refocusing of the nuclear magnetization due to a sequence of pulses, as discussed in Section I.B.5, but arise in this case from a refocusing of the magnetization due to the spinning of the solid sample. When the FID is Fourier transformed to obtain the spectrum, these echoes give rise to spinning sidebands separated from the central isotropic spectral line at frequency intervals equal to the spinning frequency f_r.

In the preceding section we showed that MAS can lead to a significant narrowing of a broadline spectrum due to CSA, irrespective of whether or not the sample is a single crystal, a powder, or amorphous. We will now be somewhat more detailed in considering a powder sample. Fig. IV.C.1 gives a coordinate system for a particular crystallite in a powdered sample. The ω coordinate frame is fixed in the rotor system. The magnetic field direction (the z direction in the laboratory frame) makes the magic angle with respect to the z coordinate in the rotor system; x, y, and z are the principal axes of the CSA tensor in the particular crystallite in question. α and β are Euler angles relating the crystallite system to the rotor system. The expression for the FID of a powder sample in the MAS experiment is (see Appendix A3)

$$T(t) = (8\pi^2)^{-1} \int_0^{2\pi} \int_0^{\pi} \int_0^{2\pi} \exp\left[i \int_0^t \omega(t')\, dt' \right] d\alpha \sin \beta \, d\beta \, d\gamma \qquad \text{(IV.C.2)}$$

where an average over a random distribution of Euler angles describing a powder with random crystallite orientation is indicated.

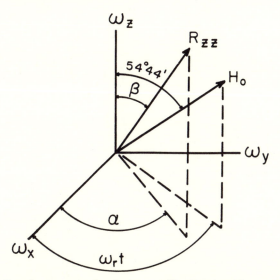

Figure IV.C.1   Coordinate system for single crystallite. Reprinted by permission from the American Institute for Physics. *J. Chem. Physics*, **76**:6 (1982), Munowitz and Griffin.

Insight is provided into the MAS experiment on powder samples by considering individual terms in Eq. IV.C.1 corresponding to particular crystallites. For example Fig. IV.C.2 gives the trajectories in the x–y plane of the magnetization associated with several individual crystallites during one rotor period. (Note that this is the x–y plane in the coordinate system rotating at the Larmor frequency around the z-axis in the laboratory frame, *not* the rotor frame fixed with the spinning sample). In each case, the magnetization begins on the x-axis of the laboratory rotating frame, and returns there at the end of the period of the rotor. By each trajectory, there is given the MAS spectrum of that particular crystallite. Fig. IV.C.3 shows the trajectory of the total magnetization (obtained by integrating over all crystallite orientations). The (rosette) pattern in the x–y plane shows the train of rotational spin echoes occurring at each point (maximum). As we have remarked earlier, an FID showing that train of echoes Fourier transforms into the narrowed spectrum with sidebands appearing as in Fig. IV.C.4.

It has been shown through an analysis of Eq. IV.C.2 that the sideband *intensities* are related to the principal values of the CSA tensor, and graphs developed to assist the experimenter in making the analysis (see Fig. IV.C.5). One first identifies the intensity of the Nth sideband I_N by varying the rotor frequency and identifying I_N as the line that moves by a frequency interval $N f_r$. One then takes the ratio I_N/I_0, where I_0 is the unshifted isotropic line, and identifies the correct curve on the appropriate graph appearing in Fig. IV.C.5. The abscissa and ordinate, μ and ρ, are defined by

$$\mu = (\gamma B)(\sigma_{33} - \sigma_{11})/\omega_r \qquad\qquad \text{(IV.C.3)}$$

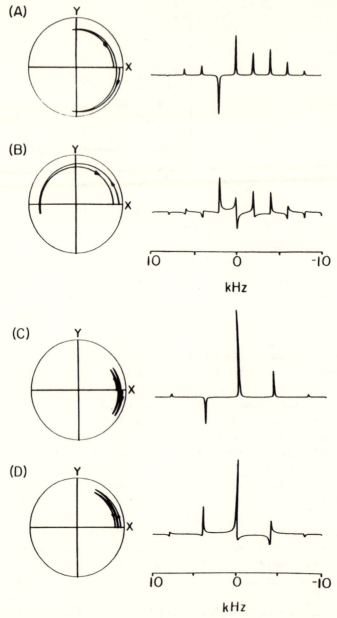

Figure IV.C.2 Trajectories for individual crystallites. Reprinted by permission from the American Institute of Physics. *J. Chem. Physics*, **81**:11 (1984), Olejniczak, Vega, and Griffin.

and

$$\rho = (\sigma_{11} + \sigma_{33} - 2\sigma_{22})/(\sigma_{33} - \sigma_{11}), \qquad \text{(IV.C.4)}$$

where we have taken the three principal values of the chemical shielding tensor components in the order

$$\sigma_{33} > \sigma_{22} > \sigma_{11}. \qquad \text{(IV.C.5)}$$

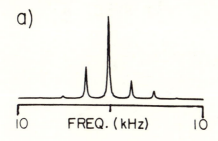

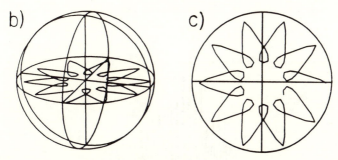

Figure IV.C.3 Trajectory for total magnetization, b); resulting spectrum, a); x–y projection, only, c). Source: Griffin, private communication.

One then superimposes the graphs in Fig. IV.C.5 and identifies as many sidebands as possible. Where they intersect, one can read off ρ and μ. With values of ρ and μ and the isotropic value of σ given by

$$\sigma = (\sigma_{11} + \sigma_{22} + \sigma_{33})/3, \tag{IV.C.6}$$

one can determine σ_{11}, σ_{22}, and σ_{33}. Some care must be taken in determining the signs of the quantities appearing in the calculation; the example worked through in Fig. IV.C.6, along with its caption, will clarify the procedures.

IV.C.2. Average Hamiltonian Theory

As shown in Appendix A4, the average Hamiltonian method represents the Hamiltonian of the system as a series with terms involving averages over the period of the time-dependent Hamiltonian:

$$\bar{H}(T) = \sum_{\mu} \bar{H}^{(\mu)} \tag{IV.C.7}$$

where the first two terms are

$$\bar{H}^{(0)} = (1/T) \int_0^T H(t')\, dt' \tag{IV.C.8}$$

and

$$\bar{H}^{(')} = (-i/2T) \int_0^T dt'' \int_0^{t''} dt [H(t''), H(t')]. \tag{IV.C.9}$$

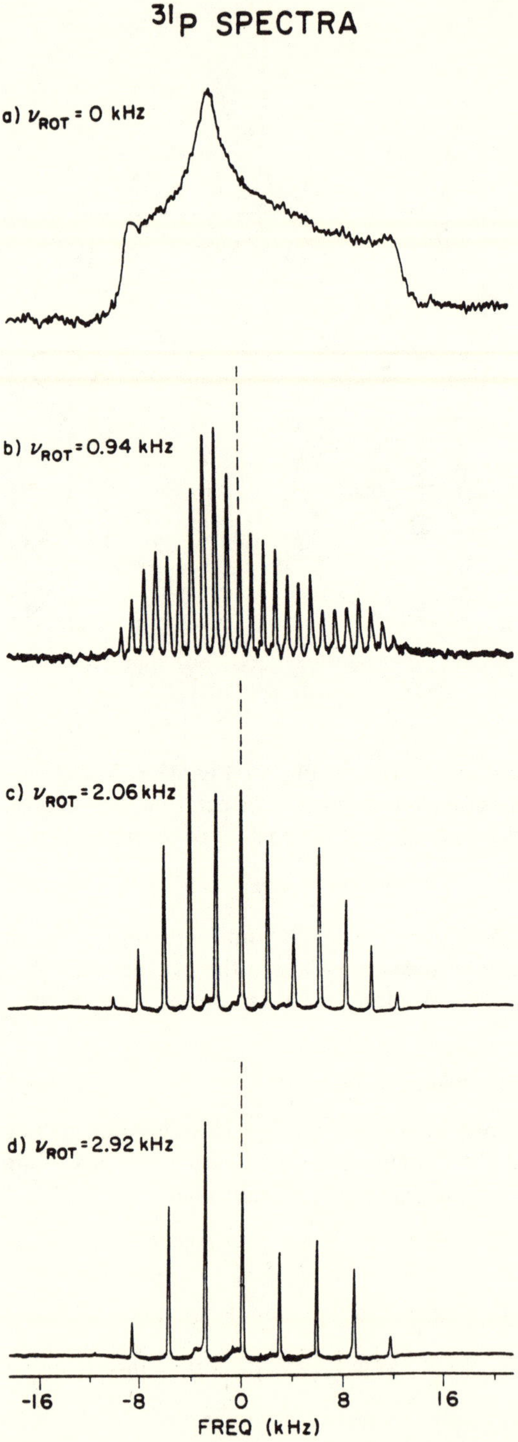

31P SPECTRA
a) ν_{ROT} = 0 kHz
b) ν_{ROT} = 0.94 kHz
c) ν_{ROT} = 2.06 kHz
d) ν_{ROT} = 2.92 kHz
-16
-8
0
8
16
FREQ (kHz)

Figure IV.C.4 Sideband patterns as a function of spinning speed. Reprinted by permission from the American Institute of Physics. *J. Chem. Physics*, **73**: 12 (1980), Hertzfeld and Berger.

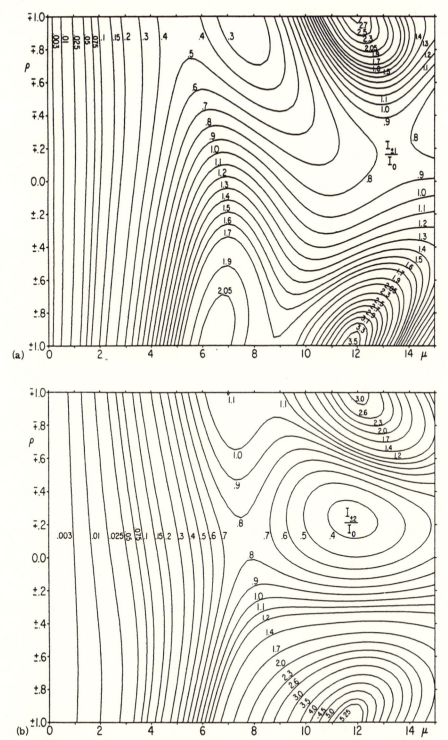

Figure IV.C.5 Contour plots of I_N/N_0 for several values of N as indicated, in the half-strip $\mu \geq 0$ and $-1 \leq \rho \leq 1$.

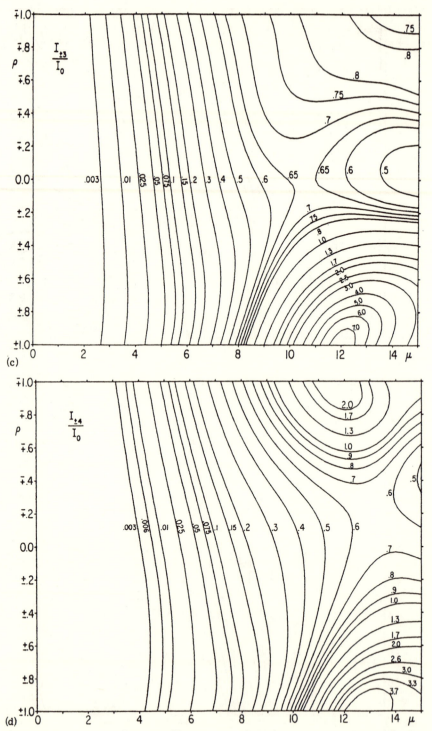

Figure IV.C.5 *(continued)* Reprinted by permission from the American Institute of Physics. *J. Chem. Physics,* **73**: 12 (1980), Hertzfeld and Berger.

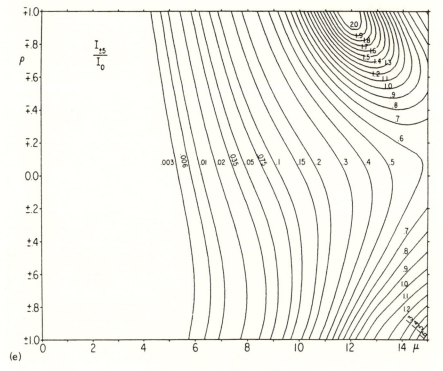

Figure IV.C.5 (continued).

The commutator appearing in Eq. IV.C.9 has the simple meaning given in Appendix A1.

Eq. IV.C.7 is a good starting point for an analysis of MAS because, as we have shown in the previous section, Hamiltonians governing systems undergoing MAS have the periodicity f_r^{-1} of the rotor rotation. It can be shown further that the magnitudes of successive terms in the expansion in Eq. IV.C.7 are related to one another by

$$\bar{H}^{(\mu+1)} \approx (\Delta/f_r)\bar{H}^{(\mu)}, \qquad (IV.C.10)$$

where Δ is a rough measure of the frequency width of the NMR spectrum due to the Hamiltonian in a non-rotating sample. One would expect that the expansion in Eq. IV.C.7 will converge only in case f_r is much greater than Δ, which is technically infeasible in certain experimental situations of interest. In fact, however, this approach does work anyway for certain types of Hamiltonian: those for which the commutator in Eq. IV.C.9 vanishes so that $H^{(1)}$ and all successive terms vanish. The commutator is equal to zero, of course, if the spin operators in the Hamiltonian commute with one another (see Appendix A1).

On examination of the chemical shift Hamiltonian H_c, given in Eq. IV.C.1, we find that it does lead to a vanishing commutator, as does the dipolar Hamiltonian for heterogeneous (unlike) nuclei. On the other hand if the nuclei are of the same type, the operators in the dipolar interaction

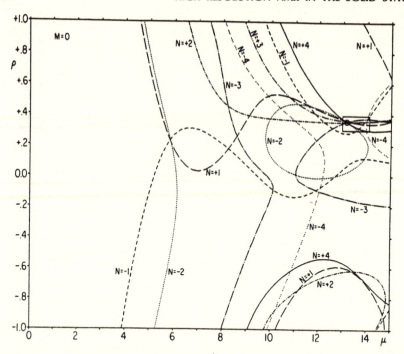

Figure IV.C.6 Graphical analysis of sideband intensities from a BDEP spectrum to obtain estimates of μ and ρ. These two parameters, combined with the isotropic chemical shift obtained from the position of the center band, give the three principal values of the chemical shift tensor. The estimated values of μ and ρ, and a generous estimate of the uncertainty in those values, are indicated respectively by the circle and rectangle at $\mu \sim 13$ and $\rho \sim 0.4$. The contours for all the sidebands except $N = -1$ and $N = -4$ pass through the circle. The relatively large deviation for $N = -1$ and $N = -4$ may be due to the fact that this region is very close to saddle points in the contour maps for these sidebands. In the vicinity of a saddle point, small errors in line intensity would produce large shifts in the corresponding contour. Notice also in this figure that the contours for $N < 0$ appear as they do in Fig. IV.C.5, while the contours for $N > 0$ appear flipped over. This is necessary in order to superimpose corresponding values of ρ (compare ordinate axis here with those in Fig. IV.C.5). Reprinted by permission from the American Institute of Physics. *J. Chem. Physics,* **73**: 12 (1980), Hertzfeld and Berger.

given in Eq. IV.B.4 do not commute. The first two interactions are examples of an *inhomogeneous* interaction, and the last is an example of a *homogeneous* interaction. According to this argument, one would expect, and indeed finds, that broadening due to an inhomogeneous interaction can be narrowed by MAS even in this case in which f_r is not large in comparison to the width of the spectrum (see Fig. IV.C.4).

Fig. IV.C.4a shows a broadline (CSA) spectrum in the absence of spinning and Figs. IV.C.4b, c, and d indicate MAS experiments at different rotation frequencies. Note that the pattern of the central isotropic line and the sidebands have envelopes that follow the broadened spectrum. On the other hand, for the case of homonuclear dipolar broadening, simple MAS at low rotation frequency is not enough to produce the desired narrowing, whereas for the CSA and to a lesser extent for the heteronuclear dipolar case (as long as the homogeneous character of the interaction imparted by homonuclear coupling and fluctuation within the other spin system is small), a narrowed pattern can be observed at achievable rotor frequencies.

IV.D. PASS, TOSS, SELTICS, AND MAGIC ANGLE HOPPING

As discussed in the previous section, one effect of MAS is the production of spinning sidebands in the NMR spectrum. Frequently the analysis of such a spectrum can lead to a confusion between isotropic absorption lines (similar to those appearing in spectra of liquid samples, which are unchanged if the rotor spinning frequency changes), and spinning sidebands separated by an integral number of rotational frequencies, Nf_r, from a central isotropic line. This problem in analysis can in principle be removed by increasing the rotor frequency, which will cause the sidebands to move away from the isotropic line, but sometimes practical considerations prevent using a high enough frequency to do the trick. Moreover, as we have seen, the sidebands themselves contain useful information. Hence there is a motivation for experimental methods for manipulating the sidebands.

This may be done by using subsidiary pulses of rf power at the resonance frequency after the initial $\pi/2$ pulse that leads to the standard FID. It can be shown (see Fig. IV.D.1) that the use of π pulses at different parts of the rotor cycle can lead to phase adjustment of spinning sidebands (PASS). For example, a π pulse used at the midpoint of a rotor period, that is at an odd half integral multiple of the period f_r^{-1}, results in a complete disappearance of the FID. Use of a π pulse at an integral multiple of the rotor period, say nf_r^{-1}, results in inverting the echo train and producing an echo at $2nf_r^{-1}$.

Use of PASS can lead to the separation of spinning bands by order (see Fig. IV.D.2a and b). One can, moreover, use a sequence of π pulses to achieve "total suppression of spinning sidebands" (TOSS), by using a sequence of π pulses at carefully determined delays.

It is clear that a π pulse can change the phase of a precessing nucleus. For example, if a single nucleus, after the initial $\pi/2$ pulse has tipped the magnetization to the x-axis, has precessed an angle π in the x–y plane, then a π pulse delivered along the x-axis will change the phase of the precessing nucleus by π. Of course, nuclei in different crystallites in a solid sample will precess at different frequencies even though they will all return to phase coincidence at the period f_r^{-1} of the rotor frequency. It can be shown that by the application of four π pulses at judiciously selected times with respect to the rotor period, one can displace in time the train of spin echoes, and hence change the phase of the spinning sidebands. It can be shown that if the time of phase coincidence (the time at which an echo appears) is displaced in time by an amount t_0, then a sideband of order N undergoes a phase change of Nt_0. This can be clarified by reference to Fig. IV.D.3. The upper trace is the original MAS spectrum without phase adjustment, with the

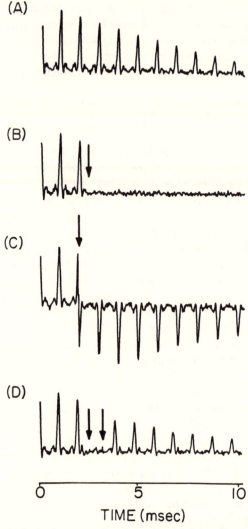

Figure IV.D.1   Effect of π pulses on train of rotational echoes. Reprinted by permission from the American Institute of Physics. *J. Chem. Physics*, **81**:11 (1984), Olejniczak, Vega, and Griffin.

sideband orders indicated. The lower trace is the PASS altered spectrum, when a *pitch* t_0 of $1/6f_r^{-1}$ has been induced. Notice the progression of phase changes in the PASS spectrum: the isotropic line is unchanged, the first order sidebands are shifted by 60° (360°/6), the second order sidebands by 120°, and the third order sidebands by 180°. The diagram to the right of the PASS spectrum makes use of the fact that a phase factor $e^{i\phi}$ can be represented as a vector with a phase angle measured between the vector and the vertical. Note that in the PASS altered spectrum, the sidebands appear at the same frequencies as in the unaltered spectrum: separated by intervals of f_r. The phases of the lines, however, are different.

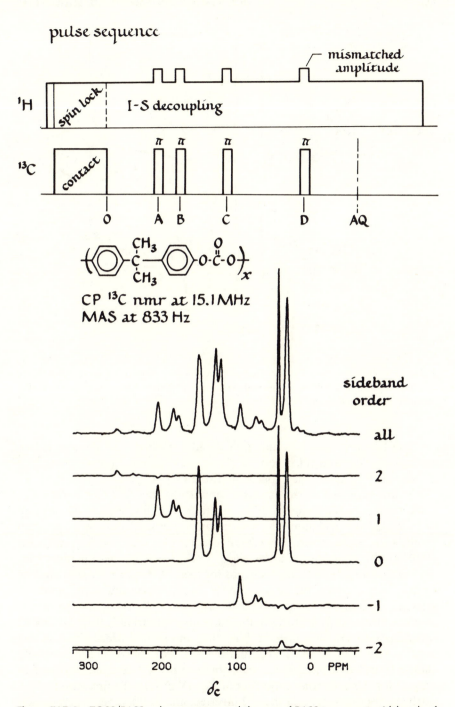

pulse sequence

Figure IV.D.2 TOSS/PASS pulse sequence and the use of PASS to separate sidebands of different order. Polycarbonate spectrum. Top curve – ordinary cross polarization experiment, 65 000 acquisitions. Lower curves – separated spectrum reconstructed from a total of 65 000 echoes. Reprinted by permission from the American Institute of Physics. *J. Chem. Physics*, **77**: 4 (1982), Dixon.

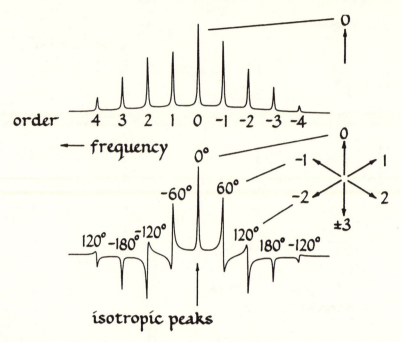

order   4 3 2 1 0 -1 -2 -3 -4

Figure IV.D.3 Phase altered spinning sidebands. Simulated spectra. Top — conventional. Bottom — typical case with PASS. Adjacent to each curve is its symbolic representation. Labeled vectors indicate phases of corresponding spinning bands. Reprinted by permission from the American Institute of Physics. *J. Chem. Physics*, **77**: 4 (1982), Dixon.

It can be shown that the timing of the four π pulses and the acquisition time (all measured in rotor periods after contact time) required to produce a given pitch t_0 are as they appear in Table IV.D.1.

We now turn to the question as to how PASS can lead to separation of a spectrum into sidebands of different order. We choose as an illustration to exemplify this technique the separation of the sidebands in a spectrum emphasizing three different sideband orders. One obtains three different PASS spectra by using pitches of 1/6, 3/6, and 5/6. The vector diagrams for first, second and third order sidebands for these three spectra appear in Fig. IV.D.4. For example, the second PASS ordered spectrum corresponds to zero and ± 2 sidebands as erect lines and the ± 1 and ± 3 order bands inverted. If one adds vectorially the diagrams appearing in Fig. IV.D.4 by superimposing the three spectra, both first and second order bands cancel out, the zero band appears erect and the ± 3 order bands are inverted. By choices of different pitches, one can obtain other PASS altered spectra to arrange for the situations appearing in Fig. IV.D.4. In the first case the $+1$ sideband appears erect and the -2 inverted, and in the other superposition the -1 band is erect and the $+2$ is inverted, with other orders missing. (It is clear that for vectors making an angle with respect to the vertical, the sideband is neither erect nor inverted but a mixture).

A similar experimental technique can be used to achieve *total suppression*

Table IV.D.1 Delays for phase altered spinning sidebands[a]

A	B	C	D	AQ	Pitch
0.0935	0.2001	0.6580	1.6357	2.1686	0.125
0.1350	0.2412	0.7096	1.6665	2.1261	0.250
0.1726	0.2682	0.7518	1.6891	2.0659	0.375
0.2098	0.2902	0.7902	1.7098	2.0000	0.500
0.2482	0.3109	0.8274	1.7318	1.9341	0.625
0.2904	0.3335	0.8650	1.7588	1.8739	0.750
0.3420	0.3643	0.9065	1.7999	1.8314	0.875
0.0772	0.1785	0.6354	1.6203	2.1724	0.0833
0.1082	0.2165	0.6770	1.6476	2.1580	0.1667
0.1350	0.2412	0.7096	1.6665	2.1261	0.2500
0.1602	0.2600	0.7383	1.6820	2.0869	0.3333
0.1850	0.2759	0.7648	1.6960	2.0442	0.4167
0.2098	0.2902	0.7902	1.7098	2.0000	0.5000
0.2352	0.3040	0.8150	1.7241	1.9558	0.5833
0.2617	0.3180	0.8398	1.7400	1.9131	0.6667
0.2904	0.3335	0.8650	1.7588	1.8739	0.7500
0.3230	0.3523	0.9819	1.7835	1.8420	0.8333
0.3646	0.3797	0.9228	1.8215	1.8276	0.9167
0.0618	0.1536	0.6123	1.6032	2.1654	0.05
0.0840	0.1880	0.6450	1.6270	2.1720	0.10
0.1024	0.2104	0.6697	1.6432	2.1628	0.15
0.1192	0.2274	0.6907	1.6558	2.1465	0.20
0.1350	0.2412	0.7096	1.6665	2.1261	0.25
0.1502	0.2530	0.7272	1.6760	2.1032	0.30
0.1652	0.2634	0.7437	1.6848	2.0786	0.35
0.1800	0.2729	0.7596	1.6933	2.0529	0.40
0.1949	0.2618	0.7751	1.7015	2.0266	0.45
0.2098	0.2902	0.7902	1.7098	2.0000	0.50
0.2249	0.2985	0.8051	1.7183	1.9734	0.55
0.2404	0.3067	0.8200	1.7271	1.9471	0.60
0.2563	0.3151	0.8348	1.7366	1.9214	0.65
0.2729	0.3240	0.8498	1.7471	1.8968	0.70
0.2904	0.3335	0.8650	1.7588	1.8739	0.75
0.3093	0.3442	0.8809	1.7726	1.8535	0.80
0.3303	0.3569	0.8976	1.7896	1.8372	0.85
0.3550	0.3730	0.9160	1.8120	1.8280	0.90
0.3877	0.3968	0.9382	1.8464	1.8346	0.95

[a] All times or angles are measured in sample revolutions after contact time. Reprinted by permission from the American Institute of Physics. *J. Chem. Physics*, **77**:4 (1982), Dixon.

of sidebands (TOSS). Under these circumstances one can use the time delays given in Table IV.D.2 in which times are measured in rotor periods after contact time. "Aq" refers to the timing of data acquisition.

A method alternative to TOSS is called "sideband elimination by temporary interruption of the chemical shift" (SELTICS). In this experimental technique one spin-locks the magnetization after the $\pi/2$ pulse that tips the magnetization into the $x-y$ plane by using the technique described in Section II.D. There are two spin–locking intervals in the course of one period of sample rotation applied at the times indicated in Fig. IV.D.5. While in TOSS,

Table IV.D.2 Delays for total sideband suppression[a]

A	B	C	D	AQ
0.2098	1.2902	. . .	. . .	2.1609
0.1885	0.2297	0.8115	1.7703	2.0000
0.1226	0.1999	0.4235	1.4668	2.2412
0.4235	0.4668	1.1226	2.1999	2.2412

[a] All times or angles are measured in sample revolutions after contact time. Reprinted by permission from the American Institute of Physics. *J. Chem. Physics*, **77**:4 (1982), Dixon.

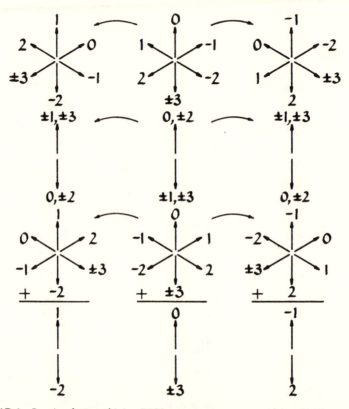

Figure IV.D.4 Routine for combining PASS spectra to separate sidebands. Three spectra with phase-adjusted spinning sidebands are combined to give a three-component spectrum separated by sideband order. Spinning band orders are indicated on each small vector. Spinning band phases are indicated by angles of labeled vectors. Reprinted by permission from the American Institute of Physics. *J. Chem. Physics*, **77**:4 (1982), Dixon.

one uses π pulses to align the magnetization from different crystallites, while in SELTICS, one turns off development due to the anisotropic part of H_C for certain intervals during the rotor period. The spin-lock power level should *not* satisfy the Hartmann–Hahn condition to prevent back-cross-polarization. Fig. IV.D.6 indicates the effectiveness of the method. The method has the merit of simplicity and is effective.

(B)

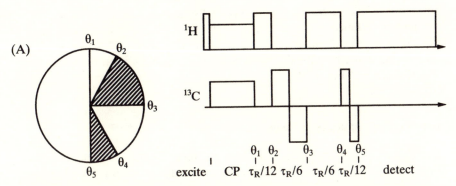

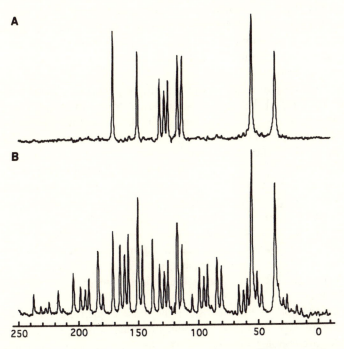

Figure IV.D.5 SELTICS pulse sequence. The $N = 6$ SELTICS sequence depicted (A) in terms of fractions of a rotor cycle and (B) as a linear sequence of pulses. Reprinted by permission from Academic Press Permissions, *J. Magn. Reson.*, Series A, **105** (1993), Hong and Harbison.

Figure IV.D.6 Use of SELTICS to eliminate spinning sidebands. ^{13}C CP-MAS spectra of tyrosine hydrochloride obtained (A) using the $N = 6$ SELTICS sequence and (B) by conventional CP-MAS. Except for the difference in pulse sequences, the spectra were obtained under identical conditions and are plotted on the same intensity scale. Experimental conditions: MAS frequency, 2.53 kHz; proton $\pi/2$ pulse, 4.5 μs; cross-polarization time, 2 ms; recycle delay, 8 s; 256 transients averaged; RF fields during SELTICS pulses, 21 kHz (^1H) and 83 kHz (^{13}C). Reprinted by permission from Academic Press Permissions, *J. Magn. Reson.*, Series A, **105** (1993), Hong and Harbison.

There is yet one more way to generate sideband free spectra. "Magic angle hopping" is a two-dimensional method in which the sample does not rotate smoothly at the magic angle but rather "hops" in 120° increments. The hopping is technically demanding, and the time required for hopping must be hidden in the t_1 (see below) evolution. Hopping in 120° or smaller increments is mathematically equivalent to continuous rotation about the magic angle (see Fig. IV.E.8).

IV.E. TWO DIMENSIONAL NMR

An experimental technique that can be very useful in simplifying the interpretation of complex NMR spectra is two dimensional (2D) NMR.

IV.E.1. Introduction

A simple example to help visualize this is the case of standard cw NMR spectroscopy in which the frequency of the transmitter is varied to obtain the spectrum of the sample while at the same time irradiating the system at a particular frequency, such as in a decoupling experiment, and obtaining spectra for a number of different decoupling frequencies. The spectra could be arranged one behind the other in a contour plot with the frequency of the spectrum along one axis and the decoupling frequency along an axis at right angles to that.

More typically one carries out a multiple pulse experiment in which one of the time intervals between pulses is varied from run to run. One can stack the spectra into an interferogram, or make a contour plot in which the variable time parameter is plotted along an axis perpendicular to the frequency axis of the Fourier-transformed FIDs. One can then take a slice perpendicular to the spectrum frequency axis parallel to the time axis. The curve along this line will be a function of time. This curve is then Fourier transformed to get a function dependent on a new frequency variable. One repeats this process for a set of slices parallel to the time axis.

One can now make a plot with the spectral frequency along what is traditionally called the f_2 axis, and along another dimension at right angles to this, called the f_1 axis, present the frequency dependence associated with variation due to changing the time parameter described earlier (see Fig. IV.E.1).

As an example one can consider the pulse sequence given in Fig. IV.E.2 in doing an experiment on a system in which two nuclei, for example ^{1}H and ^{13}C, are coupled to one another. The variable time parameter, given as τ in Fig. IV.E.2, is changed from run to run. Fig. IV.E.1 illustrates the process described in the previous paragraph. This method, using different pulse sequences, has been widely used to simplify complex spectra in the NMR spectroscopy of liquids, primarily to correlate which nuclei are coupled with which other nuclei and at what distance, and to enable easy determination of spin–spin coupling constants J.

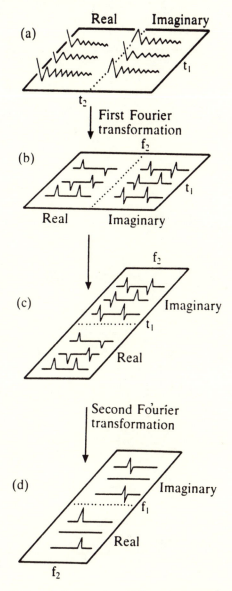

Figure IV.E.1 The 2D-FT analysis procedure. Schematic illustration of the $t_1 t_2 \rightarrow t_1 f_2 \rightarrow f_1 f_2$ double Fourier transformation process. Reprinted by permission from Oxford University Press, *Modern NMR Spectroscopy, A Guide for Chemists*, p. 97, Sanders and Hunter.

IV.E.2. DIPSHIFT

In this section we are interested in applying 2D methods in MAS experiments on solid samples. One particularly important application is to resolve the effects of the CSA interaction from the dipole interaction, using a technique called "DIPSHIFT". One uses a pulse sequence on an ^{1}H and ^{13}C system, collects numerous spectra in which a time parameter is varied, carries out

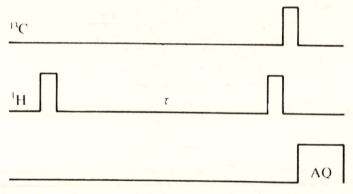

Figure IV.E.2 Basic pulse sequence for polarization transfer. Reprinted by permission from Oxford University Press, *Modern NMR Spectroscopy*, A Guide for Chemists, p. 94, Sanders and Hunter.

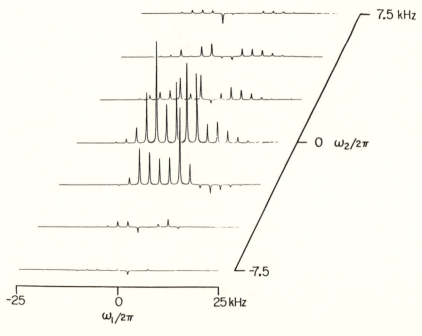

Figure IV.E.3 DIPSHIFT results. Two-dimensional chemical shift (ω_2)–dipolar (ω_1) spectrum of the model ^{13}C–^1H system. For clarity, only cross sections passing exactly through the resonances in ω_2 are shown. These are labeled by $n = \omega_2/\omega_r$, so that we may refer to dipolar slices -3 to $+3$ in moving from the near to the far end of the spectrum. Reprinted by permission from the American Institute of Physics. *J. Chem. Physics,* **76**:6 (1982), Munowitz and Griffin.

the double Fourier transform and arrives at a plot such as given in Fig. IV.E.3. One can obtain from slices parallel to the f_2 axis a pattern of sidebands whose intensities are related to the CSA parameters, and from slices parallel to the f_1 axis sidebands whose intensities can be related to the dipole interaction.

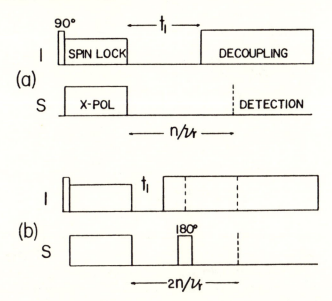

Figure IV.E.4 DIPSHIFT pulse sequences. 2D pulse sequences consisting of (1) preparation period with Hartmann–Hahn cross polarization, (2) proton coupled evolution period running from 0 to t_1, and (3) proton decoupled detection period initiated at the peak of a rotational echo. The 180° pulse in (b) is used to refocus resonance offsets in systems containing more than one isotropic chemical shift. Reprinted by permission from the American Institute of Physics. *J. Chem. Physics*, **76**:6 (1982), Munowitz and Griffin.

The DIPSHIFT experiment in its simplest form is illustrated in Fig. IV.E.4. After the cross polarization part of the experiment takes place, the heteronuclear dipolar interaction between a ^{13}C and its directly bonded protons is allowed to function for a time t_1 (t_1 is varied from run to run of the experiment and the protons are decoupled from one another during this time, using a WAHUHA sequence, or other). Then the proton is decoupled from the ^{13}C nucleus and the carbon magnetization continues to unfold under the CSA interaction, but in the absence of the heteronuclear dipolar interaction. After an integral number of rotor periods, the FID is acquired during a time t_2 for the ^{13}C magnetization.

The CSA Hamiltonian will look like Eq. IV.C.1, in which time-dependence is taken from the sample rotation. The dipolar Hamiltonian will be of the same form, but with coefficient proportional to magnetogyric ratios and inversely proportional to the cube of the internuclear separation. In analogy, the free induction decay will be

$$T(t) = e^{ip(t)} \tag{IV.E.1}$$

where $p(t)$ is the phase accumulated in the time-dependence under the influence of the Hamiltonians (see Eq. A3.3). After obtaining numerous FIDs, each with time variable t_2, for a number of different t_1, and then carrying

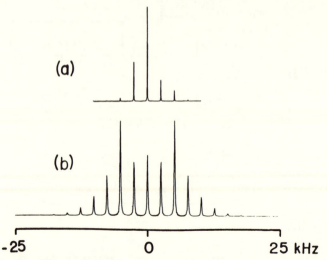

Figure IV.E.5 Two different DIPSHIFT projections. (a) on the ω_2 axis: a proton decoupled carbon spectrum; and (b) on the ω_1 axis: a proton dipolar-modulated carbon spectrum. Projections of the 2D spectrum onto the (a) ω_2 axis and (b) ω_1 axis. (a) is simply a proton decoupled spectrum, while (b) represents the response of the ^{13}C nucleus to the dipolar field of the bonded proton, all chemical shift anisotropy having been eliminated. These spectra are complete powder averages. Reprinted by permission from the American Institute of Physics. *J. Chem. Physics*, **76**:6 (1982), Munowitz and Griffin.

out Fourier transformations on both t_1 and t_2, one has

$$S(\omega_1, \omega_2) = \int \int dt_1 \, dt_2 \, e^{i\omega_1 t_1} \, e^{i\omega_2 t_2} \int \int \sin^2 \beta \, d\beta \, d\alpha$$

$$\times \cos p_1(t_1 \alpha, \beta) \exp[ip_2(t_2, \alpha, \beta)] \qquad \text{(IV.E.2)}$$

where α and β are geometrical parameters. The resulting matrix is presented as Fig. IV.E.3. Projections along the two frequency axes give the *separated* spectral structure for the heteronuclear dipolar interaction on the one hand, and the CSA sideband structure on the other (see Fig. IV.E.5).

One application of the experiment is to determine the internuclear separation between ^{13}C and the bonded proton. Internuclear distances have been accurately determined in this fashion at a level comparable to X-ray diffraction, and in a fashion that complements that technique.

IV.E.3. Dipolar Rotational Spin Echo (DRSE)

A variation of the DIPSHIFT experiment has been used extensively to study motion in solid polymers. This technique is called dipolar rotational spin echo (DRSE).

The experiment is conducted as is shown in Fig. IV.E.6. In addition to the time t_1 during which the ^{13}C-proton dipolar interaction is having an effect, and t_2 the data acquisition time, there is an optional sequence inserted

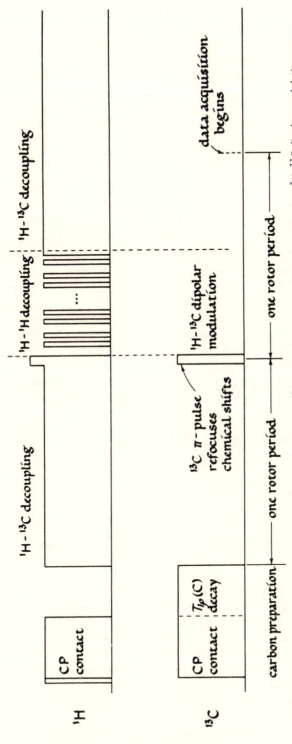

Figure IV.E.6 Pulse sequence for a dipolar rotational spin-echo ^{13}C NMR experiment. Following convention, $^{1}H–^{13}C$ dipolar modulation occurs in the time t_1, data acquisition t_2, and $T_{1\rho}(C)$ decay in t_3. Reprinted by permission from Academic Press Permissions, "Dipolar Rotational Spin-Echo ^{13}C NMR of Polymers," **124** (1982) Schaefer, McKay, Stejskal, and Dixon.

135

to select or edit the part of the ^{13}C spin system to be examined. As pictured here, there is a third time t_3 varied from experimental run to run: a time during which the carbons are spin locked without a decoupling field. The usefulness of a variable t_3 is that it allows the observer to study a mobile site in the kHz range; mobile carbons will have a short $T_{1\rho}$ and the carbons will escape the spin lock if t_3 is long enough, so that what remains in the 2D spectra represents less mobile sites. This can be observed in a comparison of the two figures given in Fig. IV.E.7 for short and long values of t_3. During the period that the heteronuclear dipolar interaction between the ^{13}C and its directly bonded proton interaction, the proton is decoupled from other protons using WAHUHA, or other homonuclear decoupling sequence, to reduce the effects of indirect coupling to other protons, via spin diffusional processes (see Section III.B).

Another change in the experiment is that low magnetic fields are used so that the 2 kHz spinning rate only gives rise to minor sidebands, so that chemical shielding does not complicate the dipolar picture. Finally, the pulse that refocuses the ^{13}C takes place before dipolar modulation; improved symmetry in the dipolar splitting and less sensitivity on the quality of the carbon π pulse result from this.

The dipolar interaction measured this way for a polymer is not exactly the same as the static dipolar interaction. First, it is scaled by the WAHUHA decoupling (see Eq. III.C.43 for an approximation to this effect), then the interaction is further reduced by motion of the molecule that is in the range of or above the strength of the interaction itself, in frequency units. The experiment therefore provides a measure of the amplitude of the motion.

IV.E.4. Other 2D Experiments with Solid Samples

Another type of experiment has been designed to gain information on partially ordered samples by the use of 2D methods. Using a relatively simple set of pulses enables one to establish a preferential axis in a partially ordered system such as a biopolymer or a synthetic polymer. The experiment, moreover, enables one to establish the orientation of that axis with the rotor axis, and how the principal-axis of the chemical shielding tensor is oriented to that preferential axis. Also, and of a particular interest, is the degree of order in the sample over a considerable range of values.

Another set of experiments has the purpose of separating the isotropic chemical shielding parameters from the anisotropic information contained in the spinning sidebands, as discussed in a previous section. If the rotor frequency is greater than the linewidth of the powder pattern due to CSA, the spinning will collapse the band to the single isotropic line, but no sideband structure is observed in this high frequency spinning limit. The object of these experiments is to obtain the isotropic information and still obtain the anisotropic information by some technique. The two methods are called "magic angle hopping" and "magic angle flipping". In the case of magic angle hopping, the static sample is tipped in three mutually orthogonal directions relative to the laboratory frame with variable times for each

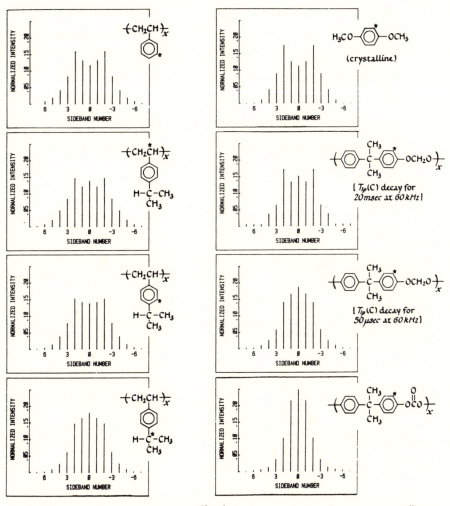

Figure IV.E.7 Dipolar Pake patterns for ^{13}C–^1H fragments of two polystyrenes, crystalline dimethoxybenzene, poly(BPA-formal), and polycarbonate. Each pattern is broken up into spinning sidebands separated by 1.894 kHz. Each spinning sideband is represented by a single point in the frequency domain which was the result of the Fourier transform of absorption-mode data. The relative intensities of the displayed sidebands are representative of the true intensities, if we make the reasonable assumption of equal widths of sidebands. Reprinted by permission from Academic Press Permissions, "Dipolar Rotational Spin-Echo ^{13}C NMR of Polymers," **127** (1982) Schaefer, McKay, Stejskal, and Dixon.

orientation (see Section IV.D). The experimental data can be displayed in two dimensions with the isotropic shift appearing on the ω_1 axis and the total powder pattern along the ω_2 axis. Magic angle flipping involves initially spinning the sample about an axis normal to $\mathbf{B}_0$, where the CSA modulation is modified but does not vanish during the evolution period, after which it is flipped to the magic angle; following this, a pulse brings the magnetization to the x–y plane to obtain the FID. An analysis of the data allows one to

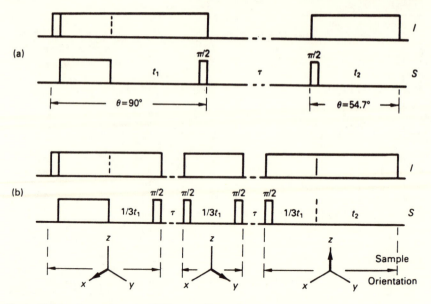

Figure IV.E.8 Separation of $\mathscr{H}_{ZS}^{aniso}$ vs. $\mathscr{H}_{ZS}^{iso}$. (a) Magic-angle "flipping" experiment: the magnetization is excited by cross polarization and precesses while the rotation axis is perpendicular to the static field, stored as longitudinal polarization while the axis of rotation is flipped in the τ-interval along the magic angle, and transferred back into transverse magnetization at the beginning of the detection period. (b) "Magic-angle hopping" experiment: a static powder is rotated along three orthogonal axes with respect to the laboratory frame, in such a way that the S-magnetization precesses for $\frac{1}{3}t_1$ in each direction. As a result, only $\mathscr{H}_{ZS}^{iso}$ is observed in ω_1, while the ω_2-domain shows the natural static power spectrum. The magnetization is temporarily stored as longitudinal polarization in the intervals τ_1 and τ_2 while the sample is rotated. Reprinted by permission from Oxford University Press, *Principles of Nuclear Magnetic Resonance in One and Two Dimensions*, **398** (1987), Ernst, Bodenhausen, and Wokaun.

separate the isotropic from the anisotropic chemical shift parameters. The details of the pulse sequences for both magic angle flipping and magic angle hopping appear as Fig. IV.E.8.

In complex systems, overlapping spinning sidebands can cause such a forest of lines in a CP/MAS experiment as to make analysis difficult or impossible. One can do a TOSS experiment (see Section IV.D) to remove the sidebands, but one loses information about the anisotropic part of the chemical shift. Several 2D NMR techniques have been proposed to make analysis simpler but also to retain anisotropic chemical shift information. In one of these, CP is followed by a TOSS sequence and a ^{13}C $\pi/2$ pulse to store one component of the magnetization along the z-axis for the variable evolution time t_1; another ^{13}C $\pi/2$ pulse and another TOSS sequence follow, and then the t_2 signal is observed. Double FT leads to a 2D plot in which sidebands are separated by order in the ω_1 dimension, and isotropic chemical shifts appear in ω_2. The sideband intensities yield the anisotropic part of the chemical shift tensor as indicated in Section IV.C (De Lacroix, 1992).

In polycrystalline solids, there is an advantage in doing NMR in zero field. This arises from the fact that since in high field, in general, only those components of local fields along the direction of the large external field are measurable, different orientations of crystallites can lead to line broadening and information is lost; information, for example, on dipolar spin coupling or the coupling between a quadrupolar nucleus ($I > 1/2$) and its local field gradient. Different zero field techniques have been developed to overcome this difficulty. In the earliest version, a sample is physically moved from a large magnetic field B_0 to a field of intermediate strength in the same direction, which is suddenly shut off. The spin system evolves for a time t_1 (varied from experiment to experiment), after which the intermediate field is restored and then the sample moved back to the high field, and a signal is obtained. The standard 2D double Fourier transformation yields the zero field spectrum, which contain the coupling information lost in the strictly high field experiment. A refinement of this method uses pulsed DC fields to detect indirectly quadrupolar nuclei. More details may be found in references Millar [1985], and Zax [1985], at the end of the chapter.

IV.F. CRAMPS

MAS is highly effective in narrowing NMR lines for sparse-spin systems (usually ^{13}C). On the other hand, in an abundant proton spin system, the linewidth, roughly 50 kHz, is sufficiently large that very high MAS speeds would be required to narrow it. Moreover, the situation is further complicated by the fact that this is homogeneous broadening (see Section I.B.6). Under these circumstances, it has been found useful to use multiple pulse techniques such as WAHUHA to reduce dipolar broadening in conjunction with MAS to reduce inhomogeneous broadening arising from chemical shift anisotropy (CSA). This technique is labeled CRAMPS, an acronym for combined rotation and multiple-pulse spectroscopy, and has been very effective in producing high resolution spectra for protons in the solid state. It can be shown that one criterion for the successful use of this method is that the rotor spinning frequency be much less than the WAHUHA sequence repetition frequency.

IV.G. DOR AND DAS

As mentioned earlier, standard MAS removes only first-order broadening due to the interaction of nuclear electric quadrupole moments with electric field gradients at their sites. Therefore, for several nuclei of solid state NMR interest that have spins greater than 1/2, such as ^{27}Al, ^{23}Na and ^{17}O, further experimental refinements were developed. One approach is called dynamic angle spinning (DAS). A single rotor is kept at an axis $\theta_1 = 37.4°$ for a time $t_1/2$, and then the axis is flipped to $\theta_2 = 79.2°$ for a further time $t_1/2 + t/2$. The usual 2D double Fourier transformation leads to a 2D plot, with the

ω_1 axis corresponding to a spectrum devoid of quadrupolar broadening. In another approach, the sample rotor is suspended inside a second rotor oriented at a different angle with respect to the external magnetic field. The technique, described as double-rotation (DOR) leads to a great increase in the solid state NMR resolution of certain quadrupolar nuclei relative to ordinary MAS. The angles of inclination of the two rotors can be shown to be the zeros of the second and fourth Legendre polynomials.

BIBLIOGRAPHY

Books

Andrew, E. R., *The Narrowing of NMR Spectra of Solids by High-speed Specimen Rotation and the Resolution of Chemical Shift and Spin Multiplet Structures for Solids*, in *Progress in Nuclear Magnetic Resonance Spectroscopy*, Editors: J. W. Emsley, J. Feeney, and L. H. Sutcliffe, Volume 8, Part 1, Pergamon Press, Oxford, 1971.

Gerstein, B. C. and Dybowski, C. R., *Transient Techniques in NMR of Solids*, Academic Press, Orlando, FL, 1985. Especially multiple pulse decoupling.

Maciel, G. E., Bronniman, C. E., and Hawkins, B. L., *High-Resolution ^1H Nuclear Magnetic Resonance in Solids via CRAMPS*, in *Advances in Magnetic Resonance*, Editor: J. S. Waugh, Volume 14, Academic Press, New York, NY, 1990.

Articles

IV.A

Kessemeier, H. and Norberg, R. E. (1967) "Pulsed Nuclear Magnetic Resonance in Rotating Solids," *Phys. Rev.* **155**, 321–337.

IV.B.1

Andrew, E. R., Bradbury, A., and Eades, R. G. (1959) "Removal of Dipolar Broadening of Nuclear Magnetic Resonance Spectra of Solids by Specimen Rotation," *Nature* **183**, 1802–1803.

Lowe, I. J. (1959) "Free Induction Decays of Rotating Solids,"*Phys. Rev. Lett.* **2**, 285–287.

Lowe, I. J. and Norbert, R. E. (1957) "Free-Induction Decays in Solids," *Phys. Rev.* **107**, 46–61.

IV.B.2

Duncan, T. M. (1990) "A Compilation of Chemical Shift Anisotropies," *The Farragut Press*, Chicago.

Facelli, J. C., Grant, D. M., and Michl, J. (1987) "Carbon-13 Shielding Tensors: Experimental and Theoretical Determination," *Acc. Chem. Res.* **20**, 152–158.

Saitô, H. (1986) "Conformation-dependent ^{13}C Chemical Shift: A New Means of Conformational Characterization as Obtained by High-resolution Solid-state ^{13}C NMR," *Magn. Reson. in Chem.* **24**, 835–852.

Stejskal, E. O., Schaefer, J., and McKay, R. A. (1977) "High-Resolution, Slow-Spinning Magic-Angle Carbon-13 NMR," *J. Magn. Reson.* **25**, 569–573.

IV.C.1

Andrew, E. R., Bradbury, A., Eades, R. G., Wynn, V. T. (1963) "Nuclear Cross-Relaxation Induced by Specimen Rotation" *Phys. Rev.* **4**, 99–100.

Herzfield, J. and Berger, A. E. (1980) "Sideband Intensities in NMR Spectra of Samples Spinning at the Magic Angle," *J. Chem. Phys.* **73**, 6021–6030.

Maricq, M. M. and Waugh, J. S. (1979) "NMR in Rotating Solids," *J. Chem. Phys.* **70**, 3300–3316.

IV.D

Dixon, W. T. (1982) "Spinning-sideband-free and Spinning-sideband-only NMR Spectra in Spinning Samples," *J. Chem. Phys.* **77**, 1800–1809.

Dixon, W. T., Schaefer, J., Sefcik, M. D., Stejskal, E. O., and McKay, R. A. (1982) "Total Suppression of Sidebands in CPMAS C-13 NMR," *J. Magn. Reson.* **49**, 341–345.

Hong, J. and Harbison, G. S. (1993) "Magic-Angle Spinning Sideband Elimination by Temporary Interruption of the Chemical Shift," *J. Magn. Reson.*, Series A, **105**, 128–136.

IV.E

Hahn, E. L. and Maxwell, D. E. (1952) "Spin Echo Measurement of Nuclear Spin Coupling in Molecules," *Phys. Rev.* **88**, 1070–1084.

Jenner, J., Meier, B. H., Bachmann, P., and Ernst, R. R. (1979) "Investigation of Exchange Processes by Two-dimensional NMR Spectroscopy," *J. Chem. Phys.* **71**, 4546–4533.

IV.E.2

Aue, W. P., Ruben, D. J., and Griffin, R. G. (1982) "Uniform Chemical Shift Scaling in Rotating Solids," *J. Magn. Reson.* **46**, 354–357.

Aue, W. P., Ruben, D. J., and Griffin, R. G. (1984) "Uniform Chemical Shift Scaling: Application to 2D Resolved NMR Spectra of Rotating Powdered Samples," *J. Chem. Phys.* **80**, 1729–1738.

Hester, R. K., Ackerman, J. L., Cross, V. R., and Waugh, J. S. (1975) "Resolved Dipolar Coupling Spectra of Dilute Nuclear Spins in Solids," *Phys. Rev. Lett.* **34**, 993–995.

Munowitz, M. G. and Griffin, R. G. (1982) "Two-dimensional Nuclear Magnetic Resonance in Rotating Solids: An Analysis of Line Shapes in Chemical Shift-dipolar Spectra," *J. Chem. Phys.* **76**, 2848–2858.

IV.E.3

Gall, C. M., DiVerdi, J. A., and Opella, S. J. (1981) "Phenylalanine Ring Dynamics by Solid-State ^{2}H NMR," *J. Am. Chem. Soc.* **103**, 5039–5043.

Schaefer, J., McKay, R. A., Stejskal, E. O., and Dixon, W. T. (1983) "Dipolar Rotational Spin-Echo ^{13}C NMR of Polymers," *J. Magn. Reson.* **52**, 123–129.

Schaefer, J., Sefcik, M. D., Stejskal, E. O., and McKay, R. A. (1981) "Separated Proton Local Fields in Polymers by Magic-Angle Carbon-13 Nuclear Magnetic Resonance," *Macromolecules* **14**, 280–283.

Schaefer, J., Stejskal, E. O., McKay, R. A., and Dixon, W. T. (1984) "Molecular Motion in Polycarbonates by Dipolar Rotational Spin-Echo C-13 NMR," *Macromolecules* **17**, 1479–1489.

Schaefer, J., Stejskal, E. O., McKay, R. A., and Dixon, W. T. (1984) "Phenylalanine Ring Dynamics by Solid-State C-13 NMR," *J. Magn. Reson.* **57**, 85–92.

Schaefer, J., Stejskal, E. O., Perchak, D., Skolnick, J., and Yaris, R. (1985) "Mechanism of the Ring-Flip Process in Polycarbonate," *Macromolecules* **18**, 368–373.

IV.E.4

De Jong, A. F., Kentgens, A. P. M., and Veeman, W. S. (1984) "Two-Dimensional Exchange NMR in Rotating Solids: A Technique to Study Very Slow Reorientations," *Chem. Phys. Lett.* **109**, 337–342.

De Lacroix, S. F., Titman, J. J., Hagemeyer, A., and Spiess, H. W. (1992) "Increased Resolution in MAS NMR Spectra by Two-Dimensional Separation of Sidebands by Order," *J. Magn. Reson.* **97**, 435–443.

Ellett Jr., J. D. and Waugh, J. S. (1969) "Chemical Shift Concertina," *J. Chem. Phys.* **51**, 2851–2858.

Harbison, G. S. and Spiess, H. W. (1986) "Two-Dimensional Magic-Angle-Spinning NMR of Partially Ordered Systems," *Chem. Phys. Lett.* **124**, 128–134.

Harbison, G. S., Raleigh, D. P., Herzfeld, J., and Griffin, R. G. (1985) "High-Field 2D Exchange Spectroscopy in Rotating Solids," *J. Magn. Reson.* **64**, 284–295.

Millar, J. M., Thayer, A. M., Bielecki, A., Zax, D. B., and Pines, A. (1985) "Zero Field NMR and NQR with Selective Pulses and Indirect Detection," *J. Chem. Phys.* **83**, 934–938.

Zax, D. B., Bielecki, A. Zilm, K. W., Pines, A., and Weitekamp, D. P. (1985) "Zero Field NMR and NQR," *J. Chem. Phys.* **83**, 4877–4905.

IV.F

Bronnimann, C. E., Hawkins, B. I., Zhang, M., and Maciel, G. E. (1988) "Combined Rotation and Multiple Pulse Spectroscopy as an Analytical Proton Nuclear Magnetic Resonance Technique for Solids," *Analytical Chem.* **60**, 1743–1750.

Burum, D. P., Linder, M., and Ernst, R. R. (1981) "Low-Power Multipulse Line Narrowing in Solid-State NMR," *J. Magn. Reson.* **44**, 173–188.

Jackson, P. and Harris, R. K. (1988) "A Practical Guide to Combined Rotation and Multiple-Pulse NMR Spectroscopy of Solids," *Magn. Reson. in Chem.* **26**, 1003–1011.

Olejniczak, E. T., Vega, S., and Griffin, R. G. (1984) "Multiple Pulse NMR in Rotating Solids," *J. Chem. Phys.* **81**, 4804–4817.

Scheler, G., Haubenreisser, U., and Rosenberger, H. (1981) "High-Resolution ^{1}H NMR in Solids with Multiple-Pulse Sequences and Magic-Angle Sample Spinning at 270 MHz," *J. Magn. Reson.* **44**, 134–144.

IV.G

Chmelka, B. F., Mueller, K. T., Pines, A., Stebbins, J., Wu, Y., and Zwanziger, J. W. (1989) "Oxygen-17 NMR in Solids by Dynamic-Angle Spinning and Double Rotation," *Nature* **339**, 42–43.

Samoson, A., Lippmaa, E., and Pines, A. (1988) "High Resolution Solid-state N.M.R. Averaging of Second-order Effects by Means of a Double-rotor," *Mol. Phys.* **65**, 1013–1018.

V

Spectrometer and Probe Design

In this chapter, we focus on several experimental aspects of the NMR process. We believe that the experimenter using NMR will profit by being aware of what is happened in the experiment even using a turn-key commercial spectrometer, in that he or she will be better prepared to cope with experimental difficulties, in addition to being in a better position to make experimental improvements.

V.A. THE BASIC SPECTROMETER

An NMR experiment involves, on the one hand, transmission of rf to the sample, and on the other hand, detection of the signal representing the response of the sample. In this chapter we will discuss modern solutions to some of the problems of transmission and detection in NMR. There will be a brief discussion of impedance matching and tuned circuits. There follows a somewhat less familiar review of the impedance properties of a transmission line. Then both single and double resonance probe circuits will be discussed. The theory of crossed diodes and the Lowe–Tarr Tee will be outlined, followed by a treatment of detectors in general, then phase detectors in more detail, culminating in a discussion of complex phase detection. Signal-to-noise (S/N) and video filters will be treated in this context, followed by a discussion of complex Fourier Transform acquisition parameters with emphasis on how to choose them. The chapter continues with discussion of experimental (electronic) artifacts and how they are dealt with by phase cycling. The basic electronic principles that we will be discussing deal with how to get the power into the sample, how to get out of the signal from the responding sample, how to reject noise from the signal, and how to assess the physical constraints of the system.

Discussions of the hardware required for MAS experiments appear in references Bartuska [1981], Doty [1981], Eckman [1980], and Zilm [1978].

V.B. PROBE CIRCUITS

In this section, we treat some of the more important aspects of probe circuit design.

V.B.1. Tuned RF Circuits

A tuned LC circuit (see Fig. V.B.1) has a resonance angular frequency given by

$$\omega_0 = (LC)^{-1/2}, \tag{V.B.1}$$

corresponding to a resonance frequency

$$f_0 = \omega_0/2\pi. \tag{V.B.2}$$

In an NMR experiment, the sample sits in the coil of a circuit tuned to a frequency at or near the resonance frequency of the nuclei of the sample in the external magnetic field. Power is fed into the tuned circuit from an rf transmitter through a transmission line. In order to maximize the power into the tuned circuit, we must consider impedance matching.

If one has a circuit with a source of impedance $Z_S = R_S + iX_S$ and terminal potential E_S terminated with a load of impedance $Z_L = R_L + iX_L$, it is straightforward to show that the average power consumed by the load is given by

$$P_{\text{av}} = R_S E_S^2/((R_S + R_L)^2 + (X_S + X_L)^2)^{1/2}. \tag{V.B.3}$$

The value of Z_L for which the load will absorb maximum power is easily determined by maximizing P_{av} with respect to R_L and X_L, and is given by

$$R_L = R_S, \qquad X_L = -X_S \tag{V.B.4}$$

To get the maximum power from the transmitter providing the rf to the sample containing the resonance nuclei, we need to match impedances. Typically, a transmission line is used to connect the two which will have a characteristic impedance of about 50 ohms which is purely resistive (see Appendix 5). For that reason, transmitters are usually designed to have 50 ohm (resistive) output impedance and we strive to achieve the same input impedance in our probe circuit. Toward this end, we will consider the case of circuits designed for independent tuning for resonance frequency and for impedance matching.

In practice, the energy losses in an inductor L can be simulated by a resistance, r_0, as indicated in Fig. V.B.2. For this circuit, we have

$$1/Z_1 = i\omega C + 1/(r_0 + i\omega L)$$
$$= r_0/(r_0^2 + \omega^2 L^2) + i(\omega C - \omega L/(r_0^2 + \omega^2 L^2)). \tag{V.B.5}$$

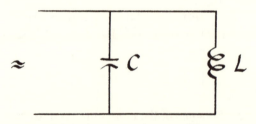

Figure V.B.1 LC circuit (parallel).

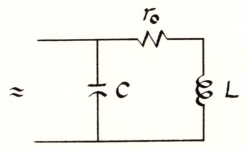

Figure V.B.2 *LC* circuit (parallel) with r_0 simulating circuit losses, primarily due to *L*.

The impedance will have no imaginary part at the resonance condition

$$\omega_0 C = \omega_0 L/(r_0^2 + \omega_0^2 L^2),\tag{V.B.6}$$

which leads to

$$\omega_0 = (LC)^{-1/2}(1 + (r_0^2/\omega_0^2 L^2))^{-1/2}.\tag{V.B.7}$$

We define the quality factor, Q, of an inductor with a series resistance r_0 as the ratio of its reactance at resonance to its resistance:

$$Q = \omega_0 L/r_0.\tag{V.B.8}$$

In typical tuned parallel circuits, $Q \gg 1$, so that

$$r_0/\omega_0 L \ll 1,\tag{V.B.9}$$

which leads to

$$\omega_0 \approx (LC)^{-1/2},\tag{V.B.10}$$

and the current through the circuit at resonance for a voltage E is, from Eqs. V.B.5, V.B.9 and V.B.10,

$$I = E/Z_1 \approx Er_0/\omega_0^2 L^2 \approx E(Cr_0/L).\tag{V.B.11}$$

A useful relation is obtained by considering a circuit with resistor R_0, capacitance C, and inductor L all in parallel (see Fig. V.B.3). It is tempting to ascribe R_0 to capacitor losses, but, in reality, losses occur throughout the circuit. The current will be

$$I = E/Z_2 = E(-i\omega C + 1/R_0 + 1/i\omega L).\tag{V.B.12}$$

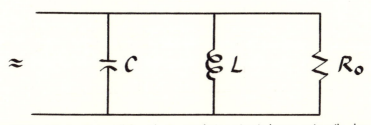

Figure V.B.3 *LC* circuit (parallel) with R_0 simulating circuit losses, primarily due to *C*.

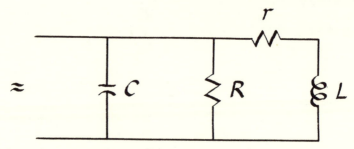

Figure V.B.4 *LC* circuit (parallel) with two resistors R and r simulating circuit losses.

If we use the resonance condition (Eq. V.B.10), we have

$$I \approx E/R_0. \qquad (V.B.13)$$

A comparison of Eqs. V.B.11 and V.B.13 indicates that, at resonance, a resistance r_0 in series with an inductor L is nearly equivalent to a shunt resistor R_0 in parallel if

$$R_0 = L/Cr_0. \qquad (V.B.14)$$

Next, let us consider the more general circuit appearing in Fig. V.B.4. We now have an additional resistance R in parallel with the circuit analogous to the ones shown as Fig. V.B.2 and V.B.3, so, at resonance, we have an impedance given by

$$1/Z = 1/R + Cr/L. \qquad (V.B.15)$$

Combining a number of the preceding equations, we obtain, for a real coil, at resonance,

$$Q \approx \omega_0 L/r_0 \approx R_0/\omega_0 L \qquad (V.B.16)$$

and

$$1/Z \approx 1/R_0 \approx 1/R + Cr/L. \qquad (V.B.17)$$

We now consider the question of impedance matching as it relates to getting the maximum power into a tuned circuit in the inductive coil of which our NMR sample will sit. In Fig. V.B.5 we have a parallel tuned circuit using

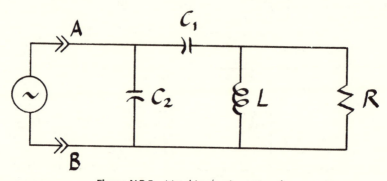

Figure V.B.5   Matching/tuning network.

two capacitors, C_1 and C_2, in series rather than a single capacitor. The utility of using two capacitors lies in the fact that tuning the probe and matching impedances can both be accomplished by varying C_1 and C_2. First, though, note that the equivalent capacitance of the series combination is

$$C_s = C_1 C_2 / (C_1 + C_2), \qquad \text{(V.B.18)}$$

from elementary considerations. The resonance frequency of the tuned circuit is simply

$$\omega_0 \approx 1/LC_s. \qquad \text{(V.B.19)}$$

We now attach a source of rf across the terminals A and B in Fig. V.B.5. This source is typically a transmission line, with a low internal impedance (for example, 50 ohms or so). In calculating the impedance Z_{in} across AB into the tuned circuit, we use the extension of Eq. V.B.5 for the three-legged circuit shown. A little algebra, combined with the approximation

$$R \gg |\omega L| \qquad \text{(V.B.20)}$$

(which follows from Eqs. V.B.8 and V.B.14 and the assumption that Q is much greater than one), leads to

$$Z_R \approx R(C_1)^2/(C_1 + C_2)^2 \qquad \text{(V.B.21)}$$

for the real part of the impedance of the network.

By varying the relative magnitudes of C_1 and C_2, we can match the impedance Z_R to the 50 ohms of the transmission line. For example, if $R = 10^4$ ohms and $L = 10^{-6}$ H, the values

$$C_1 = 35.8 \text{ pF} \qquad \text{and } C_2 = 2.72 \text{ pF} \qquad \text{(V.B.22)}$$

give a resonance angular frequency $\omega_0 = 2\pi \times 100$ MHz and a matched impedance of 50 ohms.

For higher frequencies, the circuit in Fig. V.B.5 may not work. An alternative circuit that will work in that case is shown in Fig. V.B.6.

alternative matching/tuning networks

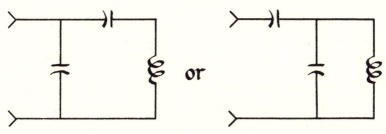

Figure V.B.6   One of these may work when the other will not.

V.B.2. Transmission Lines

We now turn to a consideration of transmission lines (see Fig. V.B.7). Not only are the properties of transmission lines important for impedance matching, they provide the circuit designer with considerable flexibility in NMR circuits. A number of modern experimental developments, to be discussed later in this chapter, depend on those properties. We will devote a reasonable amount of space to a rather careful development of these properties, which may be somewhat less familiar to the reader than standard AC circuits with discrete circuit elements.

We will assume that the line is straight coaxial cable, and that we are treating only steady state conditions by ignoring transients. To begin with, it can be shown (Appendix A5) that the characteristic impedance of a long coaxial line with no reflections is typically about 50 ohms, and is purely resistive (no reactance). We have already referred to this with regard to impedance matching.

The wavelength of an electromagnetic wave traveling on a transmission line follows directly from the standard relation between frequency f, wave velocity v, and wavelength λ:

$$v = f\lambda. \tag{V.B.23}$$

The frequency is simply the frequency of the source voltage, which, in terms of the angular frequency ω, is just

$$f = \omega/2\pi, \tag{V.B.24}$$

and the wave velocity is related to the index of refraction of the dielectric in the cable which is, in turn, just the square root of its dielectric constant:

$$v = ck^{-1/2}, \tag{V.B.25}$$

where c is the velocity of light in a vacuum, 3×10^8 m/s. For polyethylene, this gives a wave velocity of about 2×10^8 m/s.

For a long transmission line with no reflections, the line appears to be terminated by a pure resistive load with the characteristic impedance of about 50 ohms (see Appendix A5). The wave crests move to the right with velocity v, with a distance λ between successive crests (see Fig. V.B.8).

If the line is shorted at the right end, the wave will be completely reflected, which, as it interferes with the incident wave, forms a standing wave (nodes and points of maximum amplitude staying at fixed positions) with a current

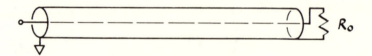

coaxial transmission line

Figure V.B.7 Schematic of coaxial transmission line terminated with its characteristic impedance.

voltages along transmission line (length = λ)

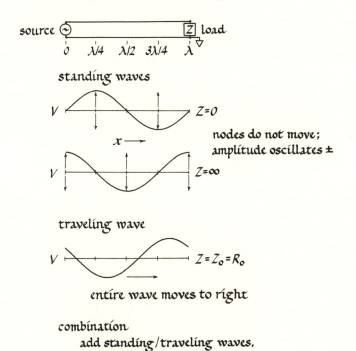

Figure V.B.8   Voltage waves along a transmission line of length corresponding to λ for the source frequency for various terminations.

maximum amplitude point at the shorted end (see Fig. V.B.8). There will be a node one-quarter of a wavelength from the shorted end. If the end of the line is open (infinite resistance – see Fig. V.B.8), there will also be complete reflection and standing waves will be set up, but with a node at the open end. Note that a shorted line has a voltage pattern equivalent to an open line at the point $\lambda/4$ from its end, and, correspondingly, an open line looks shorted at a point $\lambda/4$ from its end.

If the load at the right end of the line is resistive, but has a resistance different from the characteristic impedance, there will also be a reflected wave, but with an amplitude less than that of the incident wave, so there will be a standing wave superimposed on a traveling wave (see Fig. V.B.8).

We define a reflection coefficient ρ for a point a distance z from the source as the ratio of the amplitude of reflected wave at that point to that of the incident wave at that point. If we write the impedance at point z as

$$Z(z) = V(z)/I(z) \qquad (V.B.26)$$

and let Z_0 represent the characteristic impedance, or the impedance in the absence of any reflected wave, it can be shown (see Appendix A5), that

$$\rho = (Z/Z_0 - 1)/(Z/Z_0 + 1). \tag{V.B.27}$$

Z/Z_0 is called the normalized impedance.

Note that for the case $Z = Z_0$, the reflection coefficient is zero. Physically, this represents the fact that if the load impedance is matched to the line impedance, none of the incident power is reflected, in accordance with what we would expect.

Use of Eq. V.B.27 for the reflection coefficient leads us quickly to values for ρ for a variety of terminal load impedances Z:

1. $Z = Z_0$ (matched impedances): $\rho = 0 + i0$.
2. $Z = 0$ (short circuit): $\rho = -1 + i0$ (reflected wave undergoes 180° change of phase).
3. $Z = \infty$ (open circuit): $= 1 + i0$ (reflected wave in phase with incident wave).
4. $Z = nZ_0$ (n a positive integer): $\rho = (n - 1)/(n + 1) + i0$, $(n > 1)$.

We will now introduce the Smith chart, which provides a useful graphical method for determining the effects of various types of transmission line lengths and terminations. We begin by writing the relevant parameters in terms of real and imaginary parts:

$$\rho = u + iv \tag{V.B.28}$$

and

$$Z/Z_0 = z_n = r_n + ix_n, \tag{V.B.29}$$

where the subscript n refers to normalization of the impedance.

Eqs. V.B.27, V.B.28, and V.B.29 give

$$u + iv = (r_n + ix_n - 1)/(r_n + ix_n + 1). \tag{V.B.30}$$

Multiplying through by the denominator of Eq. V.B.30 and equating real parts and imaginary parts of the two sides of the equation lead to

$$r_n(u - 1) - x_n v = -(u + 1) \tag{V.B.31}$$

and

$$r_n v + x_n(u - 1) = -v. \tag{V.B.32}$$

The elimination of x_n and a little algebra yield

$$[u - r_n/(r_n + 1)]^2 + v^2 = 1/(r_n + 1)^2. \tag{V.B.33}$$

In a u–v Cartesian plane, this is a circle of radius $1/(r_n + 1)$ centered on the u-axis at position $r_n/(r_n + 1)$.

Fig. V.B.9 shows circles as the loci, in the u–v plane, of a set of circles corresponding to different values of r_n, the normalized resistance.

We now return to Eqs. V.B.31 and V.B.32 and solve them by eliminating r_n, rather than the normalized reactance $x_n = X/Z_0$. This time, we obtain the

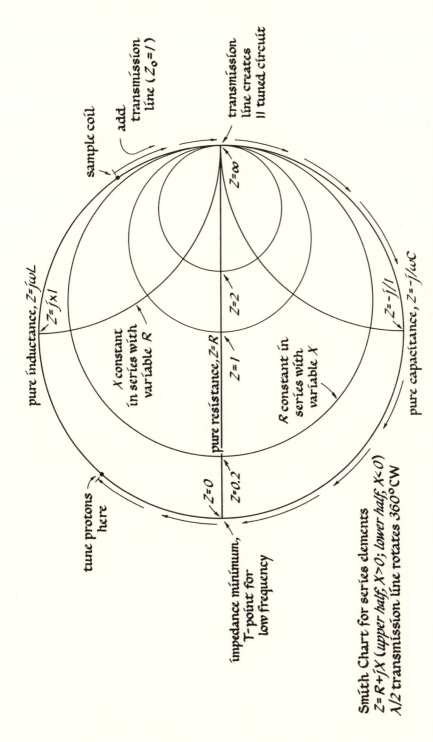

Figure V.B.9 Example of Smith chart being used to analyze McKay double-tuned probe circuit (see Section V.B.4).

Smith Chart for series elements
$Z = R + jX$ (upper half $X>0$; lower half $X<0$)
$\lambda/2$ transmission line rotates 360° CW

pure inductance, $Z=j\omega L$

$Z=j x1$

X constant in series with variable R

pure resistance, $Z=R$

R constant in series with variable X

pure capacitance, $Z=-j/\omega C$

$Z=-j/1$

$Z=2$

$Z=1$

$Z=\infty$

$Z=0.2$

$Z=0$

impedance minimum, T-point for low frequency

tune protons here

sample coil

add transmission line ($Z_o = 1$)

transmission line creates ‖ tuned circuit

following:

$$(u - 1)^2 + (v - 1/x_n)^2 = (1/x_n)^2. \qquad \text{(V.B.34)}$$

We have, again, a set of circles in the u–v plane with centers at $u = 1$ and $v = 1/x_n$, with radii $1/x_n$. The Smith chart, Fig. V.B.9, is the superposition of the two sets of circles just described.

Note that the first set of circles defines loci of points of equal resistance with variable reactances, while the second set of circles, orthogonal to the first, define loci of points of equal reactance with variable resistances, and that the bounding circle corresponds to $r_n = 0$. Moreover, the horizontal diameter corresponds to a circle from the second set with an infinite radius, or a reactance of zero. Therefore, on the Smith chart, points on this line, representing zero reactance, correspond to the total impedance simply equal to the resistance. Points in the bottom half of the chart represent negative reactances, and hence a predominantly capacitive effect; points in the upper half represent positive reactances, and hence a predominantly inductive effect.

We will now show that adding a length of transmission line to a circuit has an effect that can be determined in a simple matter from a Smith chart. It is shown in Appendix A5 that the amplitude of the voltage at a point z on a transmission line is $V = V_1 \exp(-\gamma z) + V_2 \exp(+\gamma z)$, where $\gamma = \alpha + i\beta$, where α is an attenuation coefficient and β is related to the wavelength (see Eqs. A5.7 and A5.10).

In Eq. A5.7, the first term on the right-hand side represents the incident wave, and the second term the reflected wave.

From Eq. A5.7, we see that the reflection coefficient at two points d and d_1 ($d > d_1$), are given by

$$\rho(d) = (V_2/V_1) \exp(2\gamma d) \qquad \text{(V.B.35)}$$

and

$$\rho(d_1) = (V_2/V_1) \exp(2\gamma d_1), \qquad \text{(V.B.36)}$$

so that

$$\rho(d_1) = \rho(d) \exp(2\gamma(d_1 - d)) \qquad \text{(V.B.37)}$$

Now using Eq. A5.10, we have

$$\rho(d_1) = \rho(d) \exp(2\alpha(d_1 - d)) \exp(2i\beta(d_1 - d)). \qquad \text{(V.B.38)}$$

This means that

$$|\rho(d_1)| = |\rho(d)| \exp(2\gamma(d_1 - d)) \qquad \text{(V.B.39)}$$

and the phases $\delta(d_1)$ and $\delta(d)$ are related by

$$\delta(d_1) = \delta(d) + 2\beta(d_1 - d). \qquad \text{(V.B.40)}$$

From Eq. A5.13, we have a relation between β and λ: the phase factor for a wave traveling to the right is $\exp(i(\omega t - 2\pi z/\lambda))$, which is to be compared with $\exp(i(\omega t - \beta z))$. So we have

$$\beta = 2\pi/\lambda. \qquad \text{(V.B.41)}$$

When this is compared with Eq. V.B.40, we find that adding a length of line $(d_1 - d)$ increases the phase by

$$\Delta\delta = 4\pi(d_1 - d)/\lambda. \qquad \text{(V.B.42)}$$

If $d_1 < d$, then the point at distance d_1 from the source is closer to the source than the point at a distance d, so $\Delta\delta$ is negative. Since the center of the $r_n = 0$ circle is the origin of the u–v coordinate system, a decrease in phase angle corresponds to a clockwise (since $\Delta\delta$ is negative) rotation about the bounding circle.

If $d_1 - d = n\lambda/2$, then the phase changes by $2\pi n$. If the attenuation per unit wavelength is low (small α), then adding any integral number of half-wavelengths of transmission line simply rotates a point an integral number of complete clockwise revolutions around the Smith chart, leaving you where you started. A length of transmission line equal to a quarter-wavelength corresponds to a rotation of π radians. As examples of this use of the Smith chart, consider a transmission line open at the end. This corresponds to the $Z = \infty$ point at the right end of the horizontal diameter of the $r_n = 0$ circle. Moving back a quarter of a wavelength toward the source corresponds to a clockwise rotation of π radians to the $Z = 0$ point at the other end of the horizontal diameter, corresponding physically to a short at this point, the same result we obtained earlier in a different way. If a transmission line ends in a short, the same argument leads to an infinite impedance a quarter of a wavelength back toward the source.

V.B.3. The Lowe–Tarr Tee

One device useful in the NMR pulsed experiment employs transmission lines and crossed diodes. A diode is a non-linear circuit device that passes current in only one direction (Fig. V.B.10 shows the schematic symbol for a diode, and a graph of the current I as a function of E for the device). If we define the resistance R of the device in analogy with Ohm's law,

$$R = (dI/dE)^{-1}, \qquad \text{(V.B.43)}$$

then R will be infinite or large, at least for E negative, remaining infinite through $E = 0$ to the threshold voltage, at which time it falls to a nearly constant value as E increases (see Fig. V.B.10).

If two diodes are connected in parallel but crossed (see Fig. V.B.11), then it is clear physically that the arrangement will pass current in either direction for voltages above the threshold value in either direction (see Fig. V.B.11 for the current as a function of voltage). The resistance of the combination appears as Fig. V.B.11, which indicates that crossed diodes present a large resistance to low voltages and a low resistance to high voltages.

Quarter-wave transmission lines and crossed diodes make possible the operation of the Lowe–Tarr Tee circuit for pulsed NMR experiments. The trick is to handle a high voltage pulse from the transmitter to the probe, and then, quickly, be ready to handle a low voltage nuclear induction signal from the probe. The circuit itself appears as Fig. V.B.12. When the high

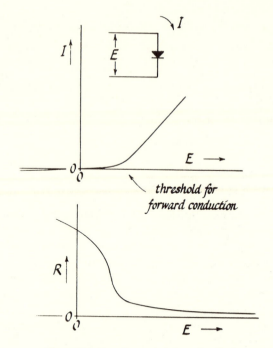

Figure V.B.10 Characteristics of a diode.

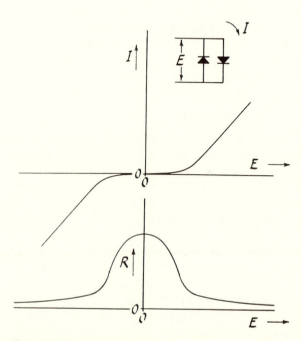

Figure V.B.11 Characteristics of a pair of crossed diodes.

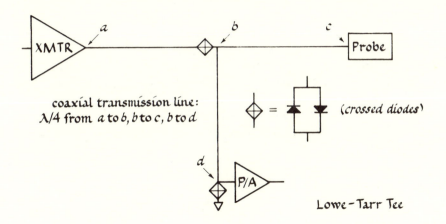

coaxial transmission line:
λ/4 from a to b, b to c, b to d

◇ = ⧩⧨ (crossed diodes)

Lowe – Tarr Tee

transmitter ON

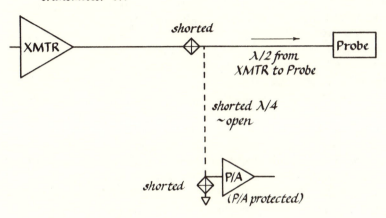

shorted

λ/2 from
XMTR to Probe

shorted λ/4
~open

shorted ◇ P/A
(P/A protected)

transmitter OFF

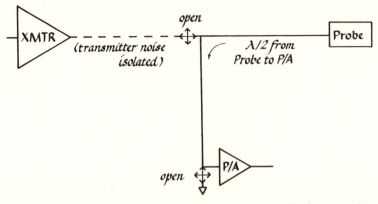

open

(transmitter noise
isolated)

λ/2 from
Probe to P/A

open

Figure V.B.12 The Lowe–Tarr Tee. Middle: how it behaves with the transmitter ON (P/A protected). Bottom: how it behaves with the transmitter OFF (XMTR noise isolated).

voltage pulse comes from the transmitter, both sets of crossed diodes provide negligible resistance, and the pulse passes through to the probe. On the other hand, the vertical quarter-wave line is shorted to ground through the conducting crossed diodes. According to our earlier arguments, a quarter of a wavelength behind the short the line behaves like an open circuit, providing a very high resistance. The pulse, therefore, does not reach the preamplifier. When the weak nuclear induction signal comes from the probe, the crossed diodes do not conduct, so that essentially the entire signal goes into the preamplifier.

V.B.4. Double Resonance Probes

Originally, double resonance experiments were done with two coils, with axes at right angles to one another, each tuned to a different frequency. This experimental arrangement is good to the extent that the two circuits were isolated electronically (there is little leakage of rf power from one source at the input of the other), but less than ideal in providing good homogeneity for both rf fields, owing to geometrical constraints. We describe here a way (due to Waugh et al.) to improve rf field homogeneity by using a single coil, while using the properties of quarter-wavelength transmission lines to preserve isolation of the two circuits.

The experimental arrangement appears as Fig. V.B.13. The single coil sits between two lengths of transmission line that are quarter-wavelength for the higher of the two frequencies: one of the lines terminates in a short, the other is open. Using the properties of quarter-wave lines, we can draw a simplified schematic of the circuit seen by the higher frequency rf (see Fig. V.B.14). This looks like a standard circuit tuned at the higher frequency.

Now consider the circuit as seen from the lower frequency source. The lengths of transmission line that are quarter-wavelength for the higher frequency are shorter than a quarter-wavelength for the lower frequency. The shorted line, then, behaves like a capacitor to the lower frequency rf, since it corresponds on a Smith chart to a clockwise rotation of less than π from the $Z = \infty$ point. Correspondingly, the other line appears as an inductor. By appropriate choice of component parameters, this

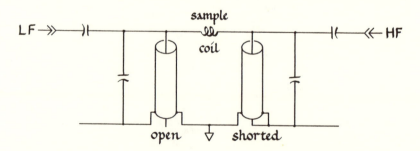

transmission lines: λ/4 at HF

Figure V.B.13 Double-tuned, single-coil probe circuit using transmission lines.

as seen by HF:

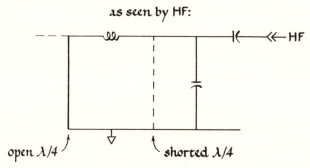

Figure V.B.14 Response of the probe to the high frequency part of the spectrometer.

as seen by LF:

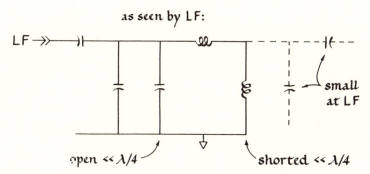

Figure V.B.15 Response of the probe to the low frequency part of the spectrometer.

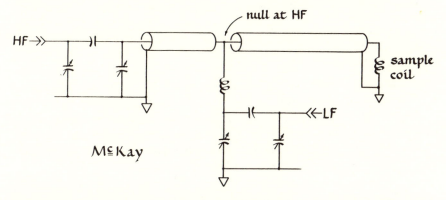

Figure V.B.16 McKay design: double-tuned, single-coil probe circuit using a single tapped transmission line. (Compare: Fig. V.B.9.).

inductor provides a low impedance path to ground, so that a schematic of the circuit as it appears to the lower frequency rf is shown as Fig. V.B.15. Thus isolation is achieved along with the rf field homogeneity made possible by using a single coil. Numerous variations of this method have been developed.

A more recent development along these lines is due to McKay (see Fig. V.B.16). The sample coil sits at the extreme right of the schematic diagram.

First consider the high frequency circuit. Just to the left of the sample coil, incoming rf sees, of course, an inductive impedance. One now adds a length of transmission line (the rightmost), designed to carry one clockwise, for high frequency rf, from "sample coil" on the Smith chart (see Fig. V.B.9) to the $Z = 0$ impedance minimum. This corresponds, then, to the "null point at HF" on the schematic diagram. A shorter length of transmission line added to the left results in a further rotation on the Smith diagram into the inductive region to the point "tune protons here". One effects this tuning with the capacitors at the leftmost part of the schematic diagram. The high frequency rf, then, moves toward the sample coil on the right, with little power diverted downward into the LF circuit, since at the null point the impedance toward the sample coil is zero. The LF part of the circuit, then, is isolated from the incoming high frequency rf.

Now consider the LF part of the circuit. Since the wavelength of the LF is longer than that of the HF, the rightmost length of transmission line provides a much smaller angle of rotation on the Smith chart from sample coil, only into the capacitive region. The high-Q inductor just below the "null point" allows one to tune for the lower frequency. When the lower frequency rf encounters the null point, it sees to the left a low capacitance and a high impedance relative to the right path to the sample coil, so most of the power will go to the sample. The HF and LF circuits, then, are effectively isolated. This method can be extended to cases of triply and higher multiply tuned coils. Note that, with the McKay circuit, only the sample coil and one end of the transmission line need be in the magnetic field. This allows for more robust tuning components, which need not be non-magnetic.

V.C. PHASE DETECTION

We will now turn our attention to detectors. The rf coil in the tuned-circuit probe of an NMR spectrometer develops, in response to a pulse of rf from the transmitter, a signal $S(t)$ from the precessing nuclei. This signal will be proportional to the x-component of the magnetization of the nuclear system, $M_x(t)$. This signal will be electronically mixed with a reference signal

$$R(t) = R_0 \cos(\omega_0 t) \qquad (V.C.1)$$

from the rf transmitter. For simplicity, take

$$S(t) = S_0 \cos(\omega_0 t + \delta). \qquad (V.C.2)$$

The output of the detector is proportional to the signal amplitude and the cosine of the phase angle:

$$T = KS_0 \cos \delta, \qquad (V.C.3)$$

where K is a constant (see Appendix A6).

One can also use a detector in which the reference signal is $\pi/2$ out of phase with the other reference signal:

$$R'(t) = R_0 \cos(\omega_0 t + \pi/2) = -R_0 \sin \omega_0 t. \qquad (V.C.4)$$

The output of this detector is

$$T' = KS_0 \sin \delta. \tag{V.C.5}$$

Using two detectors with phase difference $\pi/2$ to detect the same signal $S(t)$ is called detecting in quadrature, and has an advantage that we will see shortly.

A more realistic signal $S(t)$ from the precessing magnetization will include the effect of relaxation, and also allow for the frequency to be slightly different from the reference frequency:

$$\omega = \omega_0 + d\omega, \qquad d\omega \ll \omega_0, \tag{V.C.6}$$

so that

$$S(t) = S_0 \exp(-t/T_2) \cos([\omega_0 + d\omega]t + \delta). \tag{V.C.7}$$

We can rewrite this as follows:

$$S(t) = [S_0 \exp(-t/T_2)] \cos(\omega_0 t + [d\omega t + \delta]). \tag{V.C.8}$$

Now if the time constant θ of the detector's integration circuit is much longer than $1/\omega_0$ but much shorter than both T_2 and $1/d\omega$ (see Appendix A6), the output of the detector, in analogy with Eq. A6.5, will be

$$T = KS_0 \exp(-t/T_2) \cos(d\omega t + \delta) \tag{V.C.9}$$

and will be a slowly varying function of time, an audio frequency signal with angular frequency $d\omega$ whose amplitude is diminishing in time with a time constant T_2: the FID we discussed in Chapter I. The output of the second detector we introduced previously will be $T' = KS_0 \exp(-t/T_2) \sin(d\omega t + \delta)$. One advantage of detection in quadrature is apparent: since the cosine is an even function of its argument, detection of the in-phase signal alone does not distinguish between positive and negative values of $d\omega$. Detection in quadrature does, since the sine is an odd function of its argument.

If we now allow for a distribution of angular frequencies, this becomes

$$T(t) = \int_\infty [kS(d\omega) \exp\{-i(d\omega t + \delta) - t/T_2\}] \, d\omega, \tag{V.C.10}$$

where $S(d\omega)$ represents a distribution of angular frequencies $d\omega$ about ω_0. We can rewrite this by measuring ω about ω_0, so this becomes

$$T(t) = k \exp\{-i\delta - t/T_2\} \int_\infty S(\omega) \exp(-i\omega t) \, d\omega. \tag{V.C.11}$$

The integral looks like a kind of Fourier transform.

V.D. THE FOURIER TRANSFORM

The free induction decay (FID), the signal induced in a nuclear spin system following a $\pi/2$ pulse and written $T(t)$, is the Fourier Transform (FT) of

$F(f)$, the NMR spectrum as a function of frequency f. Mathematically,

$$F(f) = \int T(t) \exp(-2\pi i f t) \, dt, \qquad (\text{V.D.1})$$

and

$$T(t) = \int F(f) \exp(+2\pi i f t) \, df, \qquad (\text{V.D.2})$$

where both integrals are from $-\infty$ to $+\infty$. In general, $T(t)$ and $F(f)$ are complex functions. (Of course, $\omega = 2\pi f$).

In typical spectrometers, the data (measurements of $T(t)$) are acquired digitally: there will be

$$N = 2^n \qquad (\text{V.D.3})$$

pairs of data points (where n is typically in the range 7–16), corresponding to times

$$t = 0, t_s, 2t_s, \ldots, (N-1)t_s. \qquad (\text{V.D.4})$$

The FT are then calculated digitally by

$$F(j) = (1/N) \sum_{k=0}^{N-1} T(k) \exp(-i2\pi jk/N) \qquad (\text{V.D.5})$$

and

$$T(k) = \sum_{k=0}^{N-1} F(j) \exp(+i2\pi jk/N) \qquad (\text{V.D.6})$$

This is called a Discrete Fourier Transform (DFT). The individual points in the NMR spectrum calculated this way are separated by a frequency difference

$$\delta f = 1/Nt_s \qquad (\text{V.D.7})$$

The N points making up the full NMR spectrum represent a frequency width

$$\Delta f = (N-1)/Nt_s \approx 1/t_s, \qquad (\text{V.D.8})$$

centered about the phase detector reference frequency f_0.

There are several ways to adjust the spectrum window experimentally:

    a. To change the location, change the reference frequency f_0.
    b. To change the width, Δf, change the sampling interval t_s.
    c. To change the resolution, δf, without changing Δf, change the number of data pairs N by adding zeros at the end of the FID.

$F(i)$ is, in general, complex:

$$F(i) = \text{Re } F(i) + i \, \text{Im } F(i). \qquad (\text{V.D.9})$$

With an appropriate reference phase, we can make this

$$F(i) = F(i)_{\text{absorption}} + iF(i)_{\text{dispersion}}. \qquad (\text{V.D.10})$$

V.E. OPTIMIZING EXPERIMENTAL RESULTS

Signal-to-noise ratio (S/N) is improved by superimposing n different scans. The signals will add coherently, whereas the noise will add incoherently. The signal builds up linearly with the number of scans added, so that after n scans the signal S_0 increases to nS_0. The noise, however, increases only by a factor $n^{1/2}$. S/N, therefore, is given by

$$S/N = nS_0/n^{1/2}N_0 = n^{1/2}S_0/N_0. \qquad (V.E.1)$$

That is, S/N increases as the square root of the number of scans.

Apodization is a way to manipulate the data to enhance certain characteristics of the spectrum. To improve S/N, one can make use of the fact that the early part of an FID represents a higher S/N than the later part, since the signal decays with time whereas the noise level remains constant. One takes advantage of this by multiplying the FID by a function that decreases with time, such as exponentially decreasing function, to emphasize the data at the early part of the FID. On the other hand, if resolution is more of a problem than S/N, one can highlight the later part of the FID, which provides higher resolution data, by a different apodization function.

Another way to manipulate the data is to use zero-filling to increase the smoothness of the plotting. One simply fills in zero-intensity data points after the FID signal has gone into the noise, thereby removing the long time noise.

By detecting both the real and the imaginary parts of the FID, one can distinguish between positive and negative frequency differences. Fig. V.E.1 illustrates how this distinction is made experimentally.

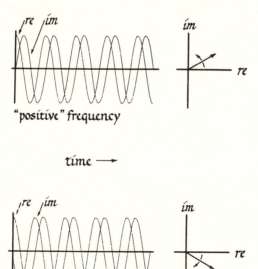

Figure V.E.1   How detection both in-phase and in-quadrature can distinguish "positive" and "negative" frequencies (relative to the phase detector frequency).

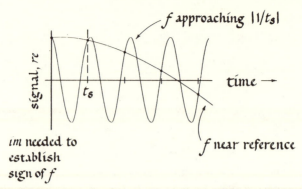

Figure V.E.2   How the discrete data collection leads to aliasing.

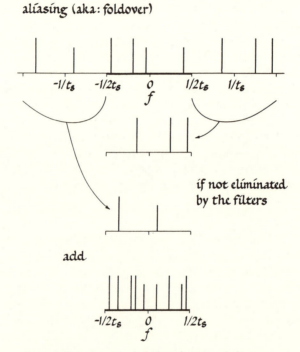

Figure V.E.3   A particularly horrific case of aliasing.

Using the discrete Fourier transform by sampling the FID only at a finite number of discrete times, leads to a certain experimental ambiguity. Fig. V.E.2 shows how a set of discrete experimental observations of the real part of two FIDs at different frequencies are indistinguishable. In fact, any FID at a frequency greater than $1/2t_s$ will be interpreted as a lower frequency – between $-1/2t_s$ and $1/2t_s$. Such a misinterpretation is described as "aliasing", and results in a fold-back of higher frequency lines into the basic frequency internal just given (see Fig. V.E.3).

A related problem involving the signal-to-noise ratio is the fold-back of

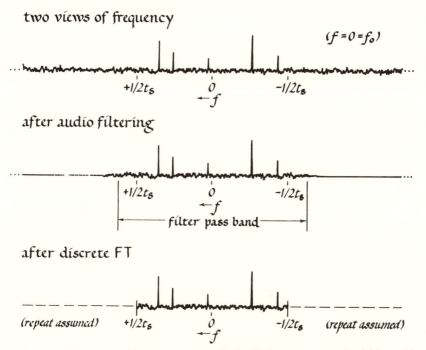

Figure V.E.4 Two views of frequency space. Audio filtering prevents noise foldover. The discrete Fourier transform treats frequency as bounded (or repeating endlessly).

high frequency noise components. A standard way to suppress this is to use audio filters (see Fig. V.E.4) to cut down noise at frequencies outside the range $-1/2t_s$ to $1/2t_s$.

Another experimental problem in handling FID signals is the deviation, by some phase angle, from the ideal case in which the real part of the FID is in phase with the transverse magnetization at $t = 0$. In fact, there is typically a phase shift, as shown in Fig. V.E.5. One can compensate for this by a phase shift factor $\exp(i\delta)$ multiplying T or F, which is equivalent experimentally to resetting the $t = 0$ position of the FID (see Fig. V.E.5). This is fine if only one frequency is present in the FID, but in all but the simplest case, more than one frequency is present, and a phase shift appropriate for one frequency will not work for another (see Fig. V.E.6). The usual way to handle this is to introduce a phase shift factor proportional to the frequency as well as the zeroth-order constant phase shift. That is, we use a multiplication factor $\exp\{i(A + Bf)\}$.

In practice we choose acquisition delays to keep $B \approx 0$ because this adjustment is, strictly speaking, only correct for our inhomogeneous distribution of frequencies. The contribution of T_2, the homogeneous part of the line shape, does not respond properly to a linear phase shift.

Experimentally, one often encounters unwanted signals that are not noise; that is, are not random, inevitable, and tending to average out upon superposition of repeated runs. These unwanted signals are collectively called

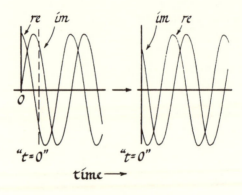

re im   im re

0

"t=0"   "t=0"

time →

delay-induced phase shift
(frequency dependent)

Figure V.E.5 How mis-setting the acquisition delay can generate an apparent (frequency dependent) phase shift.

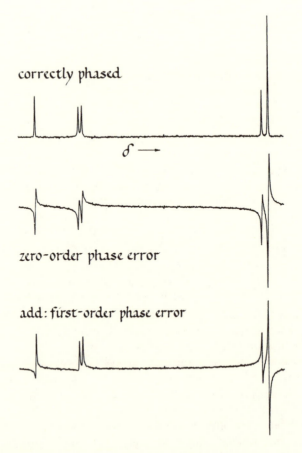

correctly phased

δ →

zero-order phase error

add: first-order phase error

Figure V.E.6 Contrasting zero-order and first-order phase shifts.

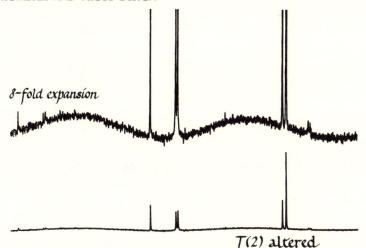

8-fold expansion

T(2) altered

Figure V.E.7   Spectrum artifact arising from single altered time-domain data point.

artifacts, and include spurious signals, distortions, pickup, pulse feed-through, probe ringing, coil ringing, sample ringing, receiver dead time, receiver gain recovery, filter transient response, detector imperfections, and so forth. We will discuss briefly a couple of these and how they may be recognized, and describe one general experimental technique, phase cycling, which is useful in suppressing a number of these artifacts.

One recognizable artifact occurs when there is a problem in picking up a point in the observation of an FID say because of receiver dead time following the rf pulse. Recall that

$$F(j) = (1/N) \sum_{k=0}^{N-1} T(k) \exp(-i2\pi jk/N). \qquad \text{(V.E.2)}$$

Missing the $T(0)$ point results in losing the constant, $k = 0$ in $F(j)$: missing the second point $T(1)$ results in losing a pair of sinusoidal curves (one each for the real and imaginary parts). Losing a point, then, results in a spectrum riding on top of a sinusoidal curve (see Fig. V.E.7), which is easy enough to recognize.

Another artifact with a recognizable signature occurs when there is slow gain recovery following the rf pulse (see Fig. V.E.8). The effect subtracts from the early, high intensity part of the FID and results in having each line in the spectrum sitting in a depression that is a broadened version of itself.

Phase cycling is an experimental technique that has the effect of canceling out a number of systematic errors and reducing artifacts. One cycles the phase of the $\pi/2$ pulses so that four successive pulses tip the magnetization (in the rotating frame) to the x-axis, the y-axis, the $-x$-axis, and the $-y$-axis. Detection in quadrature is carried out for each pulse and the FID detector outputs are appropriately combined by computer. Many systematic errors tend to cancel in this procedure.

We can summarize the workings of pulsed NMR data collection and

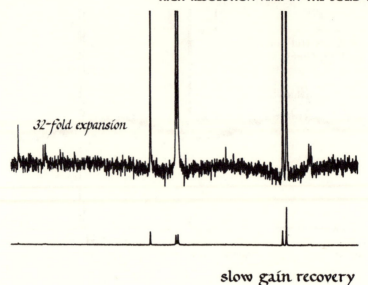

32-fold expansion

slow gain recovery

Figure V.E.8 Spectrum artifact arising when the receiver gain is slow to recover after the pulse.

display as follows:

a. Use a quadrature phase detector, and detect both real and imginary parts of $T(t)$.
b. Select the center frequency (f_0), and select the spectrum width ($-1/2t_s$ to $1/2t_s$) by selecting t_s.
c. Select the resolution ($N = 2^n$ data points per FID).
d. Set audio filters to suppress frequencies outside the spectrum width.
e. Adjust the delay before the start of data collection to cope with audio filter delays.
f. Replicate the experiment enough to obtain a desired signal-to-noise ratio.
g. Apply apodization or zero-fill or both to remove noise or increase resolution.
h. Fourier transform the results.
i. Correct phase shifts.
j. Redefine the frequency spectrum so that the reference frequency is zero and $\delta = \{(f - f_{ref})/f_0\} \times 10^6$.
k. Plot the spectrum.

BIBLIOGRAPHY

Books

Fukushima, Eiichi and Roeder, Stephen, B. W., *Experimental Pulse NMR*, Addison-Wesley, Reading, MA, 1981. Students love the approach.

Chipman, R. A., *Theory and Problems of Transmission Lines*, Shaum's Outline Series, McGraw-Hill Book Company, New York, NY, 1968.

Ellet, Jr., J. D., Gibby, M. G., Haeberlin, U., Huber, L. M., Mehring, M., Pines, A., and Waugh, J. S., *Spectrometers for Multiple-Pulse NMR*, in *Advances in Magnetic Resonance*, Editor: J. S. Waugh, Volume 5, Academic Press, New York NY, 1971.

Hayward, W. H., *Introduction to Radio Frequency Design*, Prentice-Hall, Inc., Englewood Cliffs, NJ, 1982.

Maciel, G. E., Bronniman, C. E., and Hawkins, B. L., *High-Resolution ^1H Nuclear Magnetic Resonance in Solids via CRAMPS*, in *Advances in Magnetic Resonance*, Editor: J. S. Waugh, Volume 14, Academic Press, New York, NY, 1990.

Articles

V.A

Bartuska, V. J. and Maciel, G. E. (1981) "A Magic-Angle Spinning System for Bullet-Type Rotors in Electromagnets," *J. Magn. Reson.* **42**, 312–321.

Clark, W. G. (1964) "Pulsed Nuclear Resonance Apparatus," *Rev. Sci. Instrum.* **35**, 316–333.

Clark, W. G. and McNeil, J. A. (1973) "Single Coil Series Resonant Circuit for Pulsed Nuclear Resonance," *Rev. Sci. Instrum.* **44**, 844–851.

Doty, F. D. and Ellis, P. D. (1981) "Design of High Speed Cylindrical NMR Sample Spinners," *Rev. Sci. Instrum.* **52**, 1868–1875.

Eckman, R., Alla, M., and Pines, A. (1980) "Deuterium NMR in Solids with a Cylindrical Magic Angle Sample Spinner," *J. Magn. Reson.* **41**, 440–446.

Mansfield, P. and Powles, J. G. (1963) "A Microsecond Nuclear Resonance Pulse Apparatus," *J. Sci. Instrum.* **40**, 232–238.

Zilm, K. W., Alderman, D. W., and Grant, D. M. (1978) "A High-Speed Magic Angle Spinner," *J. Magn. Reson.* **30**, 563–570.

V.B.1

Traficante, D. D. (1989) "Impedance: What It Is, and Why It Must Be Matched," *Concepts in Magn. Reson.* **1**, 73–92.

V.B.3

Lowe, I. J. and Tarr, C. E. (1968) "A Fast Recovery Probe and Receiver for Pulsed Nuclear Magnetic Resonance Spectroscopy," *J. Sci. Instrum.*, Series 2 (*J. Phys. E*) **1**, 320–322.

V.B.4

Cross, V. R., Hester, R. K., and Waugh, J. S. (1976) "Single Coil Probe with Transmission-Line Tuning for Nuclear Magnetic Double Resonance," *Rev. Sci. Instrum.* **47**, 1486–1488.

Doty, F. D., Inners, R. R., and Ellis, P. D. (1981) "A Multinuclear Double-Tuned Probe for Applications with Solids or Liquids Utilizing Lumped Tuning Elements,"*J. Magn. Reson.* **43**, 399–416.

Jiang, Y. J., Pugmire, R. J., Grant, D. M. (1987) "An Efficient Double-Tuned $^{13}C/^1H$ Probe Circuit for CP/MAS NMR and Its Importance in Linewidths," *J. Magn. Reson.* **71**, 485–494.

McKay, R. A. "Special Purpose NMR Probes," *The Encyclopedia of Nuclear Magnetic Resonance*, John Wiley & Sons, London, to be published 1995 (U.S. Patent #4,446,431, Issued May 1, 1984).

Stoll, M. E., Vega, A. J., and Vaughan, R. W. (1977) "Simple Single-Coil Double Resonance NMR Probe for Solid State Studies," *Rev. Sci. Instrum.* **48**, 800–803.

V.C

Hoult, D. I. (1978) "The NMR Receiver: A Description and Analysis of Design," *Process in NMR Spectroscopy* **12**, 41–77.

V.D

Cooley, J. W. and Tukey, J. W. (1965) "An Algorithm for the Machine Calculation of Complex Fourier Series," *Math. of Comput.* **19**, 297–301.

Ernst, R. R. and Anderson, W. A. (1966) "Application of Fourier Transform Spectroscopy to Magnetic Resonance," *Rev. Sci. Instrum.* **37**, 93–102.

Appendix A1 Vectors, Matrices, and Complex Numbers

We provide in this section a collection of definitions and procedures as a review of the mathematics of vectors, tensors, matrices, and complex numbers.

A1.A. Vectors

We will write vectors in terms of their Cartesian coordinates as

$$\mathbf{A} = A_x \mathbf{i} + A_y \mathbf{j} + A_z \mathbf{k} = A_1 \mathbf{e}_1 + A_2 \mathbf{e}_2 + A_3 \mathbf{e}_3, \tag{A1.1}$$

where $\mathbf{i}$, $\mathbf{j}$, and $\mathbf{k}$ (or $\mathbf{e}_1$, $\mathbf{e}_2$, and $\mathbf{e}_3$) are unit vectors along the 3 coordinate axes or, alternately, as a column of the components

$$\mathbf{A} = \begin{pmatrix} A_1 \\ A_2 \\ A_3 \end{pmatrix}. \tag{A1.2}$$

The *sum* of two vectors is given by

$$\mathbf{A} + \mathbf{B} = (A_1 + B_1)\mathbf{e}_1 + (A_2 + B_2)\mathbf{e}_2 + (A_3 + B_3)\mathbf{e}_3. \tag{A1.3}$$

The transpose is written

$$\tilde{\mathbf{A}} = (A_1 \quad A_2 \quad A_3). \tag{A1.4}$$

The scalar product is

$$\mathbf{A} \cdot \mathbf{B} = A_1 B_1 + A_2 B_2 + A_3 B_3 \tag{A1.5}$$

or

$$\tilde{\mathbf{A}}\mathbf{B} = (A_1 \quad A_2 \quad A_3) \begin{pmatrix} B_1 \\ B_2 \\ B_3 \end{pmatrix} = \sum_{i=1}^{3} A_i B_i. \tag{A1.6}$$

The magnitude of $\mathbf{A}$ is

$$A = (\mathbf{A} \cdot \mathbf{A})^{1/2} = (A_1^2 + A_2^2 + A_3^2)^{1/2}. \tag{A1.7}$$

The vector product of $\mathbf{A}$ and $\mathbf{B}$ has components given by

$$(\mathbf{A} \times \mathbf{B})_x = A_y B_z - A_z B_y, \tag{A1.8}$$

and cyclically.

If

$$\mathbf{C} = \mathbf{A} \times \mathbf{B}, \tag{A1.9}$$

then the magnitude of $\mathbf{C}$ is given by

$$C = AB \sin \phi \tag{A1.10}$$

where ϕ is the angle between $\mathbf{A}$ and $\mathbf{B}$. The *direction* of $\mathbf{C}$ is perpendicular to both $\mathbf{A}$ and $\mathbf{B}$, and is the direction in which the thumb of the right-hand points if the fingers curl *from* $\mathbf{A}$ *to* $\mathbf{B}$.

A1.B. Matrices and Tensors

A linear vector function T that relates one vector $\mathbf{A}$ to another vector $\mathbf{B}$,

$$\mathbf{A} = T\mathbf{B}, \tag{A1.11}$$

is called a *tensor*, and a tensor is represented by a square *matrix*:

$$M = \begin{pmatrix} M_{11} & M_{12} & M_{13} \\ M_{21} & M_{22} & M_{23} \\ M_{31} & M_{32} & M_{33} \end{pmatrix}. \tag{A1.12}$$

Matrix multiplication is defined by

$$\mathbf{M \cdot A} = \begin{pmatrix} M_{11} & M_{12} & M_{13} \\ M_{21} & M_{22} & M_{23} \\ M_{31} & M_{32} & M_{33} \end{pmatrix} \begin{pmatrix} A_1 \\ A_2 \\ A_3 \end{pmatrix} = \begin{pmatrix} V_1 \\ V_2 \\ V_3 \end{pmatrix}, \tag{A1.13}$$

where

$$V_i = \sum_{j=1}^{3} M_{ij} A_j \tag{A.14}$$

and

$$M \cdot N = \begin{pmatrix} M_{11} & M_{12} & M_{13} \\ M_{21} & M_{22} & M_{23} \\ M_{31} & M_{32} & M_{33} \end{pmatrix} \begin{pmatrix} N_{11} & N_{12} & N_{13} \\ N_{21} & N_{22} & N_{23} \\ N_{31} & N_{32} & N_{33} \end{pmatrix} = P, \tag{A1.15}$$

where

$$P_{ij} = \sum_{k=1}^{3} M_{ik} N_{kj}. \tag{A1.16}$$

We define the commutator of two matrices A and B by

$$[A, B] = AB - BA \text{ (not necessarily } = 0). \tag{A1.17}$$

A and B "commute" if

$$[A, B] = 0. \tag{A1.18}$$

The unit matrix is written

$$E = \begin{pmatrix} 1 & 0 & 0 \\ 0 & 1 & 0 \\ 0 & 0 & 1 \end{pmatrix}, \tag{A1.19}$$

and

$$ME = EM = M. \tag{A1.20}$$

The Kronecker delta is defined as

$$E_{ij} = \delta_{ij}, \tag{A1.21}$$

so that

$$\delta_{ij} = 0, \, i \neq j; \qquad \delta_{ij} = 1, \, i = j. \tag{A1.22}$$

The inverse matrix M^{-1} has the property

$$M^{-1}M = MM^{-1} = E. \tag{A1.23}$$

Other useful definitions follow:

Transpose of M: $(\tilde{M})_{ij} = M_{ji}$;

Symmetric Matrix: $M = \tilde{M}$, $M_{ij} = M_{ji}$;

Complex Conjugate: M^*, $(M^*)_{ij} = M_{ij}^*$;

Hermitian Adjoint Matrix: $M^\dagger = \tilde{M}^*$, $(M^\dagger)_{ij} = M_{ji}^*$; $\qquad$ (A1.24)

Unitary Matrix: $M^\dagger = M^{-1}$;

Hermitian Matrix: $M^\dagger = M$;

Orthogonal Matrix: $\tilde{M} = M^{-1}$.

A vector $\mathbf{A}$ can be expressed in different Cartesian coordinate systems, for example with one rotated with respect to the first:

$$\mathbf{A} = A_1\mathbf{e}_1 + A_2\mathbf{e}_2 + A_3\mathbf{e}_3 = A_1'\mathbf{e}_1' + A_2'\mathbf{e}_2' + A_3'\mathbf{e}_3' \tag{A1.25}$$

where the orthonormality of both sets of unit vectors is expressed by

$$\mathbf{e}_i \cdot \mathbf{e}_j = \delta_{ij}, \qquad \mathbf{e}_i' \cdot \mathbf{e}_j' = \delta_{ij}. \tag{A1.26}$$

If we write

$$T_{ik} = \mathbf{e}_k' \cdot \mathbf{e}_i \tag{A1.27}$$

we can show from Eqs. A1.25 and A1.26 that

$$\mathbf{A} = T\mathbf{A}'. \tag{A1.28}$$

Multiplying on the left by T^{-1}, we have

$$T^{-1}\mathbf{A} = \mathbf{A}' \tag{A1.29}$$

$$A_i = \sum_{k=1}^{3} (\mathbf{e}_k' \cdot \mathbf{e}_i)A_k' = \sum_{k=1}^{3} T_{ik}A_k'. \tag{A1.30}$$

Since

$$\mathbf{e}_1' = (\mathbf{e}_1' \cdot \mathbf{e}_1)\mathbf{e}_1 + (\mathbf{e}_1' \cdot \mathbf{e}_2)\mathbf{e}_2 + (\mathbf{e}_1' \cdot \mathbf{e}_3)\mathbf{e}_3, \tag{A1.31}$$

we can write

$$e'_1 = \alpha_1 e_1 + \beta_1 e_2 + \gamma_1 e_3, \tag{A1.32}$$

where α_1, β_1, and γ_1 are the cosines of the angles between the unit vectors. Suppose A and V are column vectors, satisfying

$$A = MV, \tag{A1.33}$$

where M is a matrix. In a different coordinate system, we have

$$A' = M'V'. \tag{A1.34}$$

But, from Eq. A1.28, we have

$$A = TA' = TM'V' = TM'T^{-1}V \text{ (since } V' = T^{-1}V)$$

$$= MV, \tag{A1.35}$$

so that

$$M = TM'T^{-1} \tag{A1.36}$$

and

$$M' = T^{-1}MT. \tag{A1.37}$$

More definitions:

$$\text{Diagonal Matrix: } D = \begin{pmatrix} d_1 & 0 & 0 \\ 0 & d_2 & 0 \\ 0 & 0 & d_3 \end{pmatrix}, \qquad D_{ij} = d_i \delta_{ij} \tag{A1.38}$$

Direct Product of two matrices:

$$A \otimes B = \begin{pmatrix} A_{11} & A_{12} \\ A_{21} & A_{22} \end{pmatrix} \otimes \begin{pmatrix} B_{11} & B_{12} \\ B_{21} & B_{22} \end{pmatrix}$$

$$= \begin{pmatrix} A_{11}B_{11} & A_{11}B_{12} & A_{12}B_{11} & A_{12}B_{12} \\ A_{11}B_{21} & A_{11}B_{22} & A_{12}B_{21} & A_{12}B_{22} \\ A_{21}B_{11} & A_{21}B_{12} & A_{22}B_{11} & A_{22}B_{12} \\ A_{21}B_{21} & A_{21}B_{22} & A_{22}B_{21} & A_{22}B_{22} \end{pmatrix}. \tag{A1.39}$$

$$\text{Trace: Tr } M = \sum_{i=1}^{3} M_{ii}. \tag{A1.40}$$

To diagonalize a matrix is to find a coordinate system with respect to which the matrix has diagonal form. Suppose M is not diagonal; we seek a coordinate transformation such that

$$M' = T^{-1}MT, \tag{A1.41}$$

where M' is diagonal; that is,

$$M'_{ij} = M'_j \delta_{ij}. \tag{A1.42}$$

Multiplying on the left by T, we obtain

$$TM' = MT. \tag{A1.43}$$

The general matrix element, then, is

$$(MT)_{ij} = \sum_k M_{ik} T_{kj} = (TM')_{ij} = \sum_k T_{ik} M'_{kj}$$

$$= \sum_k T_{ik} M'_k \delta_{kj} = T_{ij} M'_j = \sum_k T_{kj} M'_j \delta_{ik}, \tag{A1.44}$$

where we have used Eqs. A1.42 and A1.43. This leads to

$$\sum_k (M_{ik} - M'_j \delta_{ik}) T_{kj} = 0, \tag{A1.45}$$

a set of homogeneous, linear equations for the T_{kj}. A necessary and sufficient condition for a nonzero solution to exist is that the determinant of the coefficients equals zero. Solving the secular equation that arises from this procedure determines the diagonal elements M'_j, and the diagonalizing transformation T.

A1.C. The Pauli Matrices

Examples that will be encountered frequently in our discussion of NMR are:

$$\text{Vectors: } v(1/2) = \begin{pmatrix} 1 \\ 0 \end{pmatrix} \qquad v(-1/2) = \begin{pmatrix} 0 \\ 1 \end{pmatrix}, \tag{A1.46}$$

representing "spin up" and "spin down" states for a spin 1/2 nucleus, and

$$\text{Matrices: } \sigma_0 = \begin{pmatrix} 1 & 0 \\ 0 & 1 \end{pmatrix} \qquad \sigma_x = \begin{pmatrix} 0 & 1 \\ 1 & 0 \end{pmatrix}$$

$$\sigma_y = \begin{pmatrix} 0 & -i \\ i & 0 \end{pmatrix} \qquad \sigma_z = \begin{pmatrix} 1 & 0 \\ 0 & -1 \end{pmatrix}, \tag{A1.47}$$

which are the unit matrix and the Pauli spin matrices; operators which, when multiplied by $\hbar/2$, represent the Cartesian components of the spin of a spin 1/2 nucleus. Note that the matrices are Hermitian *and* unitary.

We have

$$\boldsymbol{\sigma} = \sigma_x \mathbf{i} + \sigma_y \mathbf{j} + \sigma_z \mathbf{k}. \tag{A1.48}$$

One shows easily that

$$\sigma^2 = \sigma_x^2 + \sigma_y^2 + \sigma_z^2 = 3\sigma_0. \tag{A1.49}$$

Other common notation:

$$\left. \begin{array}{l} v(1/2) = |1/2\rangle = |\alpha\rangle = \alpha \\ v(-1/2) = |-1/2\rangle = |\beta\rangle = \beta. \end{array} \right\} \tag{A1.50}$$

As treated in Section I.C.2, we will write the spin operator $\mathbf{I}$ in terms of the Pauli spin matrices as

$$I = \tfrac{1}{2}\sigma. \tag{A1.51}$$

A standard bit of notation is to define

$$I_\pm = I_x \pm iI_y. \tag{A1.52}$$

I_+ and I_- are called "raising" and "lowering" operators, since, as can be easily shown,

$$I_+|-\tfrac{1}{2}\rangle = |+\tfrac{1}{2}\rangle \tag{A1.53}$$

and

$$I_-|+\tfrac{1}{2}\rangle = |-\tfrac{1}{2}\rangle. \tag{A1.54}$$

A1.D. Spherical Polar Coordinates and Rotations

Often, spherical polar coordinates are useful. The transformation between $(x, y, \text{and } z)$ and $(r, \theta, \text{and } \phi)$ is given by

$$x = r \sin \theta \cos \phi \tag{A1.55}$$

$$y = r \sin \theta \sin \phi \tag{A1.56}$$

$$z = r \cos \theta. \tag{A1.57}$$

Of particular importance is the description of a vector $\mathbf{r}$ executing rotation with constant angular velocity ω about the z axis with unchanged length and at a constant polar angle θ. Reference to Fig. A1.1 leads to

$$|d\mathbf{r}| = (r \sin \theta)\, d\phi, \tag{A1.58}$$

and the direction of $d\mathbf{r}$ is perpendicular to both $\mathbf{r}$ and $\mathbf{k}$.
Therefore, from Eqs. A1.9 and A1.10, we can write

$$\left.\begin{aligned} \mathbf{dr} &= (\mathbf{k} \times \mathbf{r})\, d\phi \\ \frac{d\mathbf{r}}{dt} &= \left(\frac{d\phi}{dt}\mathbf{k}\right) \times \mathbf{r}. \end{aligned}\right\} \tag{A1.59}$$

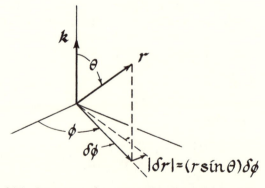

Figure A1.1 Precession of a vector of fixed length about the z-axis.

That is,

$$\frac{d\mathbf{r}}{dt} = \omega \times \mathbf{r} \tag{A1.60}$$

where

$$\omega = \frac{d\phi}{dt}\mathbf{k}. \tag{A1.61}$$

We now consider expressing the components of a vector $\mathbf{A}$ with respect to a set of coordinate axes that are themselves changing with time. For the time rate of change of $\mathbf{A}$, we have

$$\frac{d\mathbf{A}}{dt} = \left(\frac{\partial A_x}{\partial t}\mathbf{i} + \frac{\partial A_y}{\partial t}\mathbf{j} + \frac{\partial A_z}{\partial t}\mathbf{k}\right) + \left(A_x\frac{\partial \mathbf{i}}{\partial t} + A_y\frac{\partial \mathbf{j}}{\partial t} + A_z\frac{\partial \mathbf{k}}{\partial t}\right). \tag{A1.62}$$

Now from Eq. A1.56, if the coordinate system rotates with angular velocity ω,

$$\frac{\partial \mathbf{i}}{dt} = \omega \times \mathbf{i}, \tag{A1.63}$$

and so forth, so

$$\left(\frac{\partial \mathbf{A}}{dt}\right)_{\text{fixed}} = \left(\frac{\partial \mathbf{A}}{dt}\right)_{\text{rotation}} + \omega \times \mathbf{A}. \tag{A1.64}$$

A1.E. Complex Numbers

Here, for review and reference, are the basic facts about complex numbers. There is a clear relation between a complex number in the complex plane, and a vector, expressed in polar coordinates, lying in a plane. A complex number z can be written

$$z = x + iy \tag{A1.65}$$

where x and y are real numbers and $i = \sqrt{-1}$; or

$$z = r\,e^{i\theta}, \tag{A1.66}$$

where r is real and non-negative and θ is real. It is easily shown that

$$r = (x^2 + y^2)^{1/2}, \quad \theta = \tan^{-1}\left(\frac{y}{x}\right), \quad x = r\cos\theta, \quad y = r\sin\theta. \tag{A1.67}$$

Also,

$$e^{i\theta} = \cos\theta + i\sin\theta, \quad \cos\theta = \tfrac{1}{2}(e^{i\theta} + e^{-i\theta}), \quad \sin\theta = \frac{1}{2i}(e^{i\theta} - e^{-i\theta}) \tag{A1.68}$$

$$z^* = x - iy = r\,e^{-i\theta}; \quad zz^* = r^2. \tag{A1.69}$$

Appendix A2   Geometric Considerations

First note that the average value of $\cos^2 \theta$, where θ is the polar angle in a spherical polar coordinate system is,

$$\overline{\cos^2 \theta} = (1/4\pi) \int_0^{4\pi} \cos^2 \theta \, d\Omega$$

$$= (1/4\pi) \int_0^{2\pi} d\phi \int_0^{\pi} \cos^2 \theta \sin \theta \, d\theta$$

$$= 1/3. \tag{A2.1}$$

This is not to be confused with the average value of the function $\cos^2 x$ over a cycle in which all values of x between 0 and π are equally likely, for which we have

$$\overline{\cos^2 x} = (1/2\pi) \int_0^{2\pi} \cos^2 x \, dx$$

$$= 1/2. \tag{A2.2}$$

The difference in the first case is that values of θ are weighted by the $\sin \theta$ term.

We have from Eq. IV.B.11

$$\cos^2 \theta_{ij}(t) = \cos^2 \beta \cos^2 \beta_{ij} + 2 \cos \beta \cos \beta'_{ij} \sin \beta \sin \beta'_{ij} \cos(\omega_r t + \phi'_{ij})$$

$$+ \sin^2 \beta \sin^2 \beta'_{ij} \cos^2(\omega_r t + \phi'_{ij}). \tag{A2.3}$$

We now average over time. Since the average value of $\cos x$ is zero and $\cos^2 x$ is one-half (see Eq. A2.2), we have

$$\overline{\cos^2 \theta_{ij}} = \cos^2 \beta \cos^2 \beta'_{ij} + 1/2 \sin^2 \beta \sin^2 \beta'_{ij}$$

$$= \cos^2 \beta \cos^2 \beta'_{ij} + (1 - \cos^2 \beta)(1 - \cos^2 \beta'_{ij})/2$$

$$= (3 \cos^2 \beta \cos^2 \beta'_{ij} - \cos^2 \beta - \cos^2 \beta'_{ij} + 1)/2$$

$$= (3 \cos^2 \beta - 1)(3 \cos^2 \beta'_{ij} - 1)/6 + 1/3. \tag{A2.4}$$

Appendix A3   Spinning Sidebands

An elegant derivation of the production of spinning sidebands by rotating a solid sample is due to Maricq and Waugh. Associated with the Hamiltonian H_c, given in Eq. IV.C.1 and which receives a time-dependence through MAS, is a precessional angular frequency

$$\omega(t) = A + B\xi(t), \tag{A3.1}$$

where A and B are time-independent, or at least slowly varying with time, and which contain the parameters describing the precession nucleus in question, and where $\xi(t)$ is periodic with the frequency f_r of the rotor. The nuclear magnetization, initially along the x-axis in the rotating frame after a $\pi/2$ pulse, will precess in a time t with the time-dependent angular frequency given in Eq. A3.1 to an azimuthal angle

$$\phi(t) = At + B \int_0^t \xi(t')\, dt'. \tag{A3.2}$$

The FID due to this precession is proportional to $\exp[i\phi(t)]$; that is

$$T(t) = C \exp(iAt) \exp\left[iB \int_0^t \xi(t')\, dt' \right]. \tag{A3.3}$$

Since $\xi(t')$ is periodic, the second exponential will be unity for all times $t = pf_r^{-1}$, where p is any integer.

This means that $T(t)$ will consist of a sequence of echoes separated by a period f_r^{-1}. Each echo will have the same shape $g_E(t)$. The second exponential term in Eq. A3.3, then, can be written

$$\exp\left[iB \int_{?0}^t \xi(t')\, dt' \right] = \sum_p \delta(f_r t - p)^{*g_E(t)}, \tag{A3.4}$$

where we have used the delta function with the properties

$$\int_{-\infty}^{\infty} \delta(x)\, dx = 1; \qquad \delta(x) = 0, \, x \neq 0. \tag{A3.5}$$

The symbol * is used to represent convolution, defined by

$$f(x)^*g(x) = \int_{-\infty}^{\infty} f(u)g(x - u)\, du. \tag{A3.6}$$

The first exponential factor in Eq. A3.3 will be just the isotropic liquid FID for the nuclear magnetization. We have, then, for $T(t)$, a sequence of echoes separated by f_r^{-1} multiplied by the liquid state FID:

$$T(t) = T_I(t)[\Theta(f_r t)^*T_E(t)], \tag{A3.7}$$

where we have written

$$\Theta(f_r t) = \sum_p \delta(f_r t - p). \tag{A3.8}$$

The advantage of introducing this nomenclature is that the Fourier transform of $\Theta(x)$ is $\Theta(s)$, as can be seen from writing the sum of delta functions as a sum of limits of appropriately normalized gaussian functions. In view of this and other properties of Fourier transforms, we can write the NMR spectrum for the MAS case as the Fourier transform of $T(t)$:

$$F(f) = f_r F_I(f)^* \left[\Theta\left(\frac{f}{f_r}\right) F_E(f) \right] \tag{A3.9}$$

where the term in square brackets represents a sequence of spinning sidebands, separated by a frequency internal f_r, the rotational frequency of the sample.

In deriving this fundamental result, we have used that the Fourier transform of $T_1(t)[T_2(t)^*T_3(t)]$ is $F_1(f)^*[F_2(f)F_3(f)]$, and that the Fourier transform of $T(at)$ is $|a|^{-1}F(f/a)$.

Appendix A4 Average Hamiltonian Theory

From our discussion of density matrix theory (see Section I.C), we have

$$d\sigma/dt = \frac{-i}{\hbar}[H(t), \sigma]. \tag{A4.1}$$

If we assume that in the time interval from $t = 0$ to $t = T$, the Hamiltonian is piecewise constant in a set of successive small time intervals $\tau_1, \tau_2, \ldots, \tau_k,$ $\ldots, \tau_n$, and has the value H_k in the kth interval, then we can integrate Eq. A4.1 to

$$\sigma(T) = U(T)\sigma(0)U(T)^{-1}, \tag{A4.2}$$

where

$$U(T) = \exp\left(\frac{-i}{\hbar} H_n \tau_n\right) \cdots \exp\left(\frac{-i}{\hbar} H_1 \tau_1\right), \tag{A4.3}$$

and

$$T = \sum_{k=1}^{n} \tau_k. \tag{A4.4}$$

We can show that Eq. A4.3 can be written as

$$U(T) = \exp\left[\frac{-i}{\hbar} \bar{H}(T)T\right] \tag{A4.5}$$

where $\bar{H}(T)$ is called the *average Hamiltonian*. If $H(t)$ is periodic with period T, then $\bar{H}$ can also describe the time evolution of the system provided observations occur only at times periodic with period T.

Using the relation

$$e^B e^A = \exp\{A + B + [B, A]/2 + ([B, [B, A]] + [[B, A], A])/12 + \cdots\} \tag{A4.6}$$

for operators A and B that do not necessarily commute (which can be established by writing the exponentials in the product as a power series), we can write $\bar{H}(T)$ as

$$\bar{H}(T) = \bar{H}^{(0)} + \bar{H}^{(1)} + \bar{H}^{(2)} + \cdots, \tag{A4.7}$$

where

$$\bar{H}^{(0)} = \left(\sum_k H_k \tau_k\right)\bigg/ T \tag{A4.8}$$

$$\bar{H}^{(1)} = \frac{-i}{\hbar} \{[H_2\tau_2, H_1\tau_1] + [H_3\tau_3, H_1\tau_1] + [H_3\tau_3, H_2\tau_2] + \ldots, \tag{A4.9}$$

and so forth.

This generalizes to the continuous case as follows:

$$\bar{H}^{(0)} = (1/T) \int_0^T H(t')\, dt', \tag{A4.10}$$

$$\bar{H}^{(1)} = (-i/2T) \int_0^T dt'' \int_0^t dt'[H(t''), H(t')]. \tag{A4.11}$$

Appendix A5   Transmission Lines

Fig. A5.1 represents a portion of a coaxial line in terms of distributed elements: R and L are the series resistance and inductance per unit length, C is the parallel capacitance per unit length, and G is the parallel conductance (the reciprocal of the parallel resistance) per unit length. Fig. A5.2 indicates schematically the source impedance Z_s and the source voltage V_s, the current and voltage $I(z)$ and $V(z)$ a distance z from the source, and the terminal amplitude impedance Z_t. Consideration of an elemental length Δz of the line leads to the two complex valued equations

$$V(z + \Delta z) - V(z) = -R\Delta z I(z) - i\omega L\Delta z I(z) \tag{A5.1}$$

and

$$I(z + \Delta z) - I(z) = -iC\Delta z V(z) - G\Delta z V(z), \tag{A5.2}$$

where ω is the angular frequency of the source voltage. If we set the left-hand sides of Eqs. A5.1 and A5.2 to ΔV and ΔI respectively, divide through by Δz, and let Δz approach zero, we obtain

$$dV/dz = -(R + i\omega L)I \tag{A5.3}$$

and

$$dI/dz = -(G + i\omega C)V. \tag{A5.4}$$

Differentiating these two coupled equations with respect to z and solving them simultaneously leads to

$$d^2V/dz^2 - (G + i\omega C)(R + i\omega L)V = 0 \tag{A5.5}$$

and

$$d^2I/dz^2 - (G + i\omega C)(R + i\omega L)I = 0 \tag{A5.6}$$

It is easily seen by substitution that the solutions of these differential equations are

$$V = V_1 \exp(-\gamma z) + V_2 \exp(+\gamma z) \tag{A5.7}$$

and

$$I = I_1 \exp(-\gamma z) + I_2 \exp(+\gamma z), \tag{A5.8}$$

where V_1, V_2, I_1, and I_2 are complex valued and independent of z, and where

$$\gamma = [(R + i\omega L)(G + i\omega C)]^{1/2}. \tag{A5.9}$$

Now V_1, V_2, I_1, and I_2, while not functions of z, the distance along the transmission line from the source, will vary harmonically in time at the

coaxial transmission line

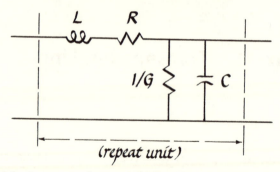

Figure A.5.1   A segment of transmission line viewed as distributed elements.

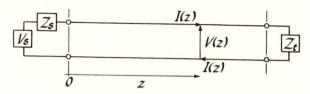

Figure A.5.2   Parameters used to describe propagation of current along a transmission line.

angular frequency ω of the source voltage: that is, each will be a constant complex number multiplied by $\exp(i\omega t)$.

If we write

$$\gamma = \alpha + i\beta, \tag{A5.10}$$

$$V_1 = V_1' \exp(i\omega t), \tag{A5.11}$$

and

$$V_2 = V_2' \exp(i\omega t), \tag{A5.12}$$

then Eq. A5.7 becomes

$$V = V_1' \exp(-\alpha z)\exp[i(\omega t - \beta z)] + V_2' \exp(+\alpha z)\exp[i(\omega + \beta z)], \tag{A5.13}$$

where the first term represents a wave moving to the right and the second term represents a wave moving to the left.

To determine the transmission line's characteristic impedance Z_0, we will assume that there are no reflected waves on the line, and drop the second terms in Eqs. A5.7 and A5.8.

Hence,

$$V = V_1 \exp(-\alpha z)\exp(-i\beta z) \tag{A5.14}$$

and

$$I = I_1 \exp(-\alpha z)\exp(-i\beta z). \tag{A5.15}$$

If we differentiate Eq. A5.13 with respect to z and substitute the result into Eq. A5.3, we obtain

$$V_1/I_1 = (R + i\omega L)/(\alpha + i\beta). \tag{A5.16}$$

Using Eq. A5.9, we obtain

$$Z_0 = V_1/I_1 = [(R + i\omega L)/(G + i\omega C)]^{1/2}, \tag{A5.17}$$

where this equation defines Z_0, the characteristic impedance of the transmission line.

From Eqs. A5.3 and A5.7, we have

$$I = [\gamma/(R + i\omega L)](V_1 \exp(-\gamma z) - V_2 \exp(\gamma z)). \tag{A5.18}$$

From Eqs. A5.8 and A5.17, we obtain

$$\begin{aligned} I &= I_1 \exp(-\gamma z) + I_2 \exp(+\gamma z) \\ &= (V_1 \exp(-\gamma z) + V_2 \exp(+\gamma z))/Z_0. \end{aligned} \tag{A5.19}$$

By definition, the impedance Z at z is simply

$$\begin{aligned} Z &= V(z)/I(z) \\ &= Z_0[V_1 \exp(-\gamma z) + V_2 \exp(+\gamma z)]/[V_1 \exp(-\gamma z) - V_2 \exp(+\gamma z)]. \end{aligned} \tag{A5.20}$$

As defined in the chapter, the reflection coefficient is the ratio of amplitude of the reflected wave to the incident wave:

$$\begin{aligned} \rho &= V_2 \exp(+\gamma z)/V_1 \exp(-\gamma z) \\ &= (V_2/V_1) \exp(2\gamma z). \end{aligned} \tag{A5.21}$$

If we divide all four terms in Eq. A5.20 by $V_1 \exp(-\gamma z)$ and use Eq. A5.21, we arrive at

$$Z/Z_0 = (1 + \rho)/(1 - \rho), \tag{A5.22}$$

or, alternatively,

$$\rho = (Z/Z_0 - 1)/(Z/Z_0 + 1). \tag{A5.23}$$

In the typical transmission line case that we will be using, we will have

$$R \ll \omega L \tag{A5.24}$$

and

$$G \ll \omega C \tag{A5.25}$$

so that Eq. A5.17 for the characteristic impedance becomes

$$Z_0 \approx (L/C)^{1/2}. \tag{A5.26}$$

Note that this is a real quantity implying that Z_0 is purely resistive.

We will evaluate this characteristic impedance for a straight coaxial transmission line with inner conductor radius a and outer conductor radius b with a material of dielectric constant k filling the space between the conductors (see Fig. A5.3).

In order to use Eq. A5.26 to get Z_0, we need the capacitance per unit length, C, and the inductance per unit length, L. To calculate the first of these, assume a uniform charge per unit length ρ on the surface of the inner conductor and $-\rho$ on the outer conductor. Gauss's law can be used to show that the electric field between the conductors a radial distance r from the

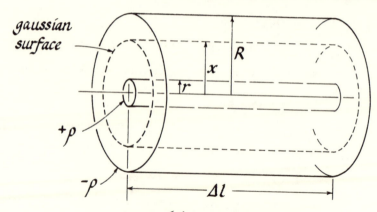

gaussian surface

+ρ

−ρ

coaxial transmission line capacitance calculation

Figure A.5.3 Figure to accompany calculation of capacitance per unit length along transmission line.

central axis of the cable is

$$E = \rho/2\pi\varepsilon r \qquad (A5.27)$$

where ε is the electric permittivity of the dielectric material. (To do this, use a cylindrical gaussian surface of unit length and radius r concentric with the axis of the cable.)

Integrating the negative of the electric field from a to b to get the potential difference between the conductors gives

$$V(b) - V(a) = (\rho/2\pi\varepsilon) \ln(b/a). \qquad (A5.28)$$

Using the definition of capacitance per unit length, we obtain

$$C = \rho/(V(b) - V(a))$$
$$= 2\pi\varepsilon/(\ln(b/a)). \qquad (A5.29)$$

Now for the inductance per unit length. Consider a current I flowing to the right (in Fig. A5.2) in the inner conductor and to the left in the outer conductor. Taking a circular path of radius x concentric with the axis and using Ampere's law, we obtain

$$B(r) = \mu I/2\pi r \qquad (A5.30)$$

for the magnetic flux density, where μ is the magnetic permeability of the dielectric material. If we consider a strip of thickness dx and length Δl (see Fig. A5.4), we find the magnetic flux through it due to the current I is

$$d(\text{flux}) = B(r)\Delta l\, dr = \mu I \Delta l\, dx/2\pi r. \qquad (A5.31)$$

If we integrate this from the surface of the inner conductor to the outer

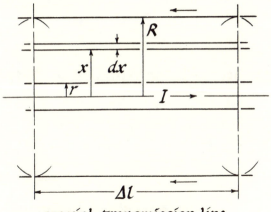

coaxial transmission line
inductance calculation

Figure A.5.4 Figure to accompany calculation of inductance per unit length along transmission line.

conductor, we get

$$\Delta(\text{flux}) = (1/2\pi)\Delta l I \mu \ln(b/a). \tag{A5.32}$$

Using the definition of inductance per unit length, we obtain

$$L = \Delta(\text{flux})/I\Delta l = (\mu/2\pi) \ln(b/a). \tag{A5.33}$$

By substituting Eqs. A5.29 and A5.33 into Eq. A5.26, we arrive at an expression for the characteristic impedance Z_0 of the transmission line:

$$Z_0 = (1/2\pi)(\mu/\varepsilon)^{1/2} \ln(b/a). \tag{A5.34}$$

But $\varepsilon = k\varepsilon_0$ and $\mu \approx \mu_0$ where ε_0 and μ_0 are the electric permittivity and magnetic permeability of free space (and k is the dielectric constant of the material between the conductors: $\mu \approx \mu_0$ is well-satisfied for such materials). Putting in standard values for μ_0 and ε_0, we obtain

$$Z_0 = (60/k^{1/2}) \ln(b/a). \tag{A5.35}$$

For typically k, b, and a, this is about 50 ohms.

Eq. A5.34 was derived for the case of a transmission line carrying a wave from left to right (coming from the source), with no reflected wave (by dropping the term representing a wave traveling from right to left). This condition is satisfied in the case of a line of infinite (in practice, very great) length.

Appendix A6   Phase Sensitive Detection

One way to make a phase sensitive detector is to multiply electronically the signal

$$S(k) = S_0 \cos(\omega_0 t + \delta) \qquad (A6.1)$$

by the reference signal

$$b(t) = b_0 \cos(\omega_0 t) \qquad (A6.2)$$

so that

$$S(t)b(t) = b_0 S_0(\cos^2 \omega_0 t \cos \delta - \cos \omega_0 t \sin \omega_0 t \sin \delta). \qquad (A6.3)$$

The detector then integrates this signal with respect to time (using, for example, a simple RC circuit) over a number of cycles; that is, for a time θ satisfying

$$\theta \gg 1/\omega_0. \qquad (A6.4)$$

The average value of $\cos^2 \omega_0 t$ over a cycle is 1/2, while the average value of $\cos \omega_0 t \sin \omega_0 t$ is zero.

Therefore we have an output for the detector of

$$T = KS_0 \cos \delta, \qquad (A6.5)$$

as presented in the text.

BIBLIOGRAPHY

Books

A3

Bracewell, R. *The Fourier Transform and its Applications*, McGraw-Hill, NY, 1965.

A5

Chipman, R. A., *Theory and Problems of Transmission Lines*, Schaum's Outline Series, McGraw-Hill Book Company, New York, NY, 1968. Answers to (mostly) qualitative questions.

Articles

A6

Hoult, D. I. (1978) "The NMR Receiver: A Description and Analysis of Design," *Progress in NMR Spectroscopy* **12**, 41–77.

Index